Perspectives in the History of Science and Technology

UNIVERSITY OF OKLAHOMA PRESS : NORMAN

Perspectives in the History of Science and Technology

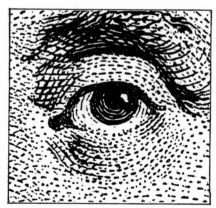 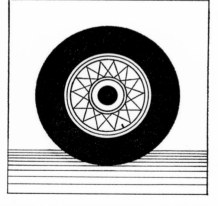

Edited by **DUANE H. D. ROLLER**

This book is published with the assistance of THE JOHN HOUCHIN FUND

International Standard Book Number: 0–8061–0952–1

Library of Congress Catalog Card Number: 77–144163

FOREWORD

The history of science is a new branch of history, a child of the twentieth century. Professionally trained historians of science are a product of the last two decades. As a consequence, this new and rapidly growing field of inquiry is as yet ill-defined.

With a few notable exceptions, nineteenth-century historians were not interested in the history of science, and the first impetus to its study came from individuals trained in one of the sciences. Some of these were teachers who felt that the history of their science offered pedagogic value in their teaching. Others were retired scientists who turned to a search for their own origins and intellectual roots. A few among these learned the techniques of the historian for study of the past; they became the founders of the field of the history of science.

The circumstances of its origin caused the history of science to be regarded as a portion of science, history being merely a tool for the study of the past of science. And knowledge of that past was largely organized in terms of the modern branches of science.

Within this general structure the tendency has been to focus attention upon the achievements of individual scientists. In part this was due to a nineteenth-century heroic view of history; in part it was because those who regarded the history of a science as a pedagogic aid in the teaching of that science wished to display its notables as exemplars to students. Reference materials are largely organized in terms of people, and the source materials for the history of science have seemed to consist almost entirely of manuscripts and publications by individual scientists. And, finally, the having of ideas is widely regarded as the prerogative of the individual.

In its earlier stages, then, the history of science tended to be accounts of the work of individual scientists of the past, grouped according to modern divisions of science.

The history of technology never has seemed as alien to historians as did the history of science. It is technology that has produced the most visible monuments of the past, and social and cultural historians have long been aware of the impact of technology upon the societies and cultures they study. To be sure, there have been written histories of individual branches of engineering and of specific types of machinery, as well as studies of the works of individual engineers. But there has never been in the historical study of technology the kind of isolation from the culture that appeared in the history of science and that tended to regard science as a body of "organized positive knowledge" whose practitioners piled new truths upon old ones in a steady progression toward obtaining total knowledge.

The second half of this century has seen drastic changes in the study of the history of science. A new breed of professional historians of science has appeared on the scene, and a considerable number of professional historians have turned their attention to the history of science as well as the history of technology. The arrival of these scholars trained in history has predictably produced changes in the history of science, for their motivations are largely historical rather than scientific. Although the source materials remain the writings of individuals, there is a growing tendency to spread beyond the confines of the writings of the scientist and to examine the culture which produced him. Although studies of individual scientists remain an essential portion of the history of science, many historians now group the results of such studies in terms of the history of an idea or a point of view or an era, rather than in terms of a twentieth-century science.

All of these aspects of the study of the history of science and technology are displayed in the papers and commentaries comprising this volume. They were originally presented at a symposium at the University of Oklahoma April 8–12, 1969, during the Inaugural Year of the ninth president of the University, J. Herbert Hollomon. The Midwest Junto and the Society for the History of Technology joined the University in sponsoring the symposium. Rather than attempt organization of the symposium around some scientific or historical "theme," it was decided to ask historians of science and technology to speak on subjects of their own choosing. Two other specialists in the same area of research prepared and presented commentaries on each paper, based upon advance copies. Eight presentations and sixteen commentaries comprised the Symposium and are now published here.

In addition, the volume contains two other papers which were given at the time of and in conjunction with the Symposium, one the

Rosetta Briegel Barton Lecture, by Professor Ihde, the other the Phi Alpha Theta Lecture of Father Clark.

These twenty-six papers and commentaries by distinguished historians of science and technology offer a cross-section of research and attitudes in these fields in the seventh decade of the twentieth century.

DUANE H. D. ROLLER

Norman, Oklahoma
March 15, 1971

CONTENTS

Perspectives in the History of Science and Technology

The Kuhnian Paradigm and the Darwinian Revolution in Natural History

By John C. Greene, The University of Connecticut

THE PUBLICATION OF Thomas Kuhn's *The Structure of Scientific Revolutions* in 1962 was an important milestone in the development of the historiography of science. It was the first attempt to construct a generalized picture of the process by which a science is born and undergoes change and development. The main stages of development envisaged by Kuhn's model may be summarized as follows:

1. A pre-paradigm stage in which the natural phenomena that later form the subject matter of a mature science are studied and explained from widely differing points of view.

2. The emergence of a paradigm, embodied in the published works of one or more great scientists, defining and exemplifying the concepts and methods of research appropriate to the study of a certain class of natural phenomena, and serving as an inspiration to further research by its promise of success in explaining those phenomena.

3. A period of normal science conducted within a conceptual and methodological framework derived from the paradigmatic achievement, involving actualization of the promise of success, further articulation of the paradigm, exploration of the possibilities within the paradigm, use of existing theory to predict facts, solving of scientific puzzles, development of new applications of theory, and the like.

4. A crisis stage of varying duration precipitated by the discovery of natural phenomena that "violate the paradigm-induced expectations that govern normal science" and marked by the invention of new theories designed to take account of the anomalous facts.

5. A relatively abrupt transition to a new paradigm brought about by the achievements of a scientific genius who defines and

exemplifies a new conceptual and methodological framework in-commensurable with the old.

    6. Continuation of normal science within the new paradigm.

Professor Kuhn's examples of the formation and transformation of paradigms are drawn entirely from the history of the physical sci-ences, but he gives us no reason to believe that his analysis is not applicable to the sciences generally. It may be worthwhile, therefore, to examine the developments leading up to the Darwinian revolution in natural history to see to what extent they fit the pattern of historical development described in Kuhn's book.

Perhaps the best way to begin the investigation is to ask: When did natural history first acquire a paradigm? When did it arrive at a state characterized by "research firmly based upon one or more past scien-tific achievements that some particular scientific community ac-knowledged for a time as supplying the foundation for its further practice"; achievements "sufficiently unprecedented to attract an en-during group of adherents away from competing modes of scientific activity," yet "sufficiently open-ended to leave all sorts of problems for the redefined group of practitioners to resolve"?

This is not an easy question to answer. On the whole, however, it seems that such a condition cannot be said to have prevailed in natural history until the emergence of systematic natural history in the late seventeenth century, its embodiment in the publications of John Ray and Joseph Pitton de Tournefort, and its apotheosis in the works of Carl Linnaeus.

Aristotle and Theophrastus had laid the foundations of scientific zoology and botany two thousand years earlier, but their achieve-ments cannot be said to have given rise to a continuing tradition of research based on their precept and example. The herbalists cannot be said to have been continuing the Theophrastian tradition, nor can Pliny, Albertus Magnus, Gesner, and Aldrovandi be said to have been the continuators of Aristotle in the same sense that Brisson, Jussieu, Candolle, Cuvier, Lamarck, Hooker, and Agassiz were con-tinuators of the tradition established by Tournefort, Ray, and Linnaeus. Doubtless the Aristotelian achievement was profounder, broader, and in some ways more fecund than that of the founders of systematic natural history, but it did not, like theirs, give rise to and dominate an enduring tradition of scientific research of the kind Kuhn has in mind when he speaks of normal science.

It may be objected, however, that systematic natural history as practiced by Ray, Tournefort, and Linnaeus, was not a science in

Kuhn's terms because it did not explain anything, but only named, classified, and described natural objects. This objection raises the difficult problem whether science can be defined in absolute terms; that is, in such a way that the definition is valid for all sciences in all periods of history.

Kuhn himself seems to favor a loose, relativistic concept of science that would allow for the fact that every great scientific revolution involves some redefinition of the nature and aim of science. He tells us that no man is a scientist unless he is "concerned to understand the world and to extend the precision and scope with which it has been ordered." On the other hand, he stresses the importance of respecting "the historical integrity of that [older] science in its own time." With respect to the acceptance and rejection of paradigms he asserts that "there is no standard higher than the assent of the relevant community," and he rejects flatly the view, which he attributes to Charles Gillispie, that "the history of science records a continuing increase in the maturity and refinement of man's conception of the nature of science."

It would seem, therefore, that whatever the ultimate truth about the nature of science may be, no objection to the scientific status of systematic natural history can be drawn from Kuhn's book. Systematic natural historians were concerned to understand the world and to extend the precision and scope with which it was ordered. They considered themselves scientists and were so considered by their contemporaries, including the physical scientists. True, they did not consider it their business as natural historians to explain the origin of species, but neither did Newton consider it his business as a natural philosopher to explain the origin of the solar system.

Like Newton, Ray and Linnaeus took for granted a static concept of nature that regarded all the structures of nature as created and wisely designed by an omnipotent God in the beginning. This assumption of the permanence and wise design of specific forms and of the basic structures of nature generally was an essential feature of the paradigm of systematic natural history, integrally related to the belief that the aim of natural history was to name, classify, and describe.

By every criterion laid down by Kuhn there was a paradigm of systematic natural history. Emerging from the scientific achievements of Ray, Tournefort, and Linnaeus, it involved commitments on all the levels—cosmological, epistemological, methodological, etc.—mentioned by Kuhn. Embodied in manuals and popularizations, articulated with increasing precision, communicated by precept and example, celebrated in prose and verse, it dominated the field of natural history for nearly two hundred years and helped to prepare the way for a far different, far more dynamic kind of natural history. To this extent, then, we can say that Kuhn's model of scientific development seems to fit fairly well with what is known concerning the emergence of systematic natural history as a science of nature.

Having established, at least to our own satisfaction, that natural history first acquired a paradigm in the Kuhnian sense through the work of Ray, Tournefort, and Linnaeus, we next inquire when this paradigm may be said to have been supplanted by a different one. Here it seems generally agreed that the publication of Charles Darwin's *Origin of Species* was the decisive event in the transition from a static, taxonomy-oriented natural history to a dynamic and causal evolutionary biology. Whatever the exact nature and causes of the Darwinian revolution, there can be little doubt that Darwin's work inaugurated a new era in the study of organic nature. Before discussing this revolution further, however, it will be well to inquire into its genesis in order to discover whether the development of natural history from Linnaeus to Darwin followed the pattern of normal science, anomaly, crisis, and paradigm invention described by Kuhn.

At the outset of this inquiry we are confronted with a phenomenon for which Kuhn's model makes no provision, namely, the appearance of a counter paradigm coeval, or nearly so, with the establishment of the static paradigm of natural history.

In the same mid-eighteenth-century years when Linnaeus was rearing the edifice of systematic natural history on foundations laid by Ray and Tournefort, the Count de Buffon was publishing his splendid *Histoire Naturelle, Générale et Particulière*, based on a profoundly different concept of natural history from that which inspired Linnaeus and his forerunners. In Linnaeus' view, the function of the natural historian was to name, classify, and describe the productions of the earth and, above all, to search for a natural method of classification. In Buffon's opinion, classifications were arbitrary human devices that played a useful but subordinate role in the main business of natural history, which was to explain the observed uniformities in nature's productions as necessary results of the operations of the hidden system of laws, elements, and forces constituting primary, active, and causative nature. Where Linnaeus saw a world of plants and animals neatly ordered and perfectly adapted to their surroundings by the wise design of an omnipotent Creator, Buffon saw a confused array of living forms, some better adapted to their environment than others, all subject to modification through changes in climate, diet, and the general circumstances of life, all threatened in one

degree or another by the activities of man and by the gradual cooling
of the earth that had spawned these beings by its own powers. While
Linnaeus described and catalogued species, genera, orders, and
classes, searching for the natural method of classification (presumably one corresponding to a pattern in the mind of the Creator),
Buffon devoted his energies to studying the processes of generation,
inheritance, and variation by which various kinds of animals had
been produced and modified.

Like Darwin a century later, Buffon investigated the history of
domesticated plants and animals and discovered the importance of
artificial selection, both conscious and unconscious, in producing
domestic races. He conducted experiments in animal hybridization.
He compared the quadrupeds of the Old World with those of the
New, and sought to understand their similarities and differences as
effects of descent with modification. He collected and compared
fossils from Europe, Asia, and America and attempted to envisage
the epochs of earth history in the light of these discoveries. He can-
vassed the literature concerning human races, man-like apes, wild
children, pigmies, and giants, and strove to portray the history of man
as a part of the wider history of nature. Finally, like Darwin, he invented a theory of pangenesis to explain the apparent facts of heredity, growth, nutrition, and modification through environmental
change.

In all of this there was much that was speculative, much that was
tentative and incomplete. But the challenge to Linnaean precept and
practice in natural history was clear and unmistakable. Buffon had
promulgated and in a great degree exemplified a new kind of natural
history—dynamic, causal, non-teleological, time-oriented, uniformitarian in principle if not always in practice, concerned with discovering the laws and mechanisms of organic change, aimed ultimately at
control of nature through an understanding of her modes of operation. The difference in outlook between the new natural history and
the old is apparent in the two following quotations:

Linnaeus: The study of natural history, simple, beautiful, and instructive, consists in the collection, arrangement, and exhibition of
the various productions of the earth.

Buffon: In general, kindred of species is one of those mysteries of
Nature, which man can never unravel, without a long continued and
difficult series of experiments. . . . Is the ass more allied to the horse
than the zebra? Does the wolf approach nearer to the dog than the
fox or jackal? At what distance from man shall we place the large

apes, who resemble him so perfectly in conformation of body? Are all
the species of animals the same now that they were originally? . . .
Have not the feeble species been destroyed by the stronger, or by
the tyranny of man. . . . Does not a race, like a mixed species, proceed
from an anomalous individual which forms the original stock? How
many questions does this subject admit of; and how few of them are
we in a condition to solve? How many facts must be discovered before
we can even form probable conjectures?[1]

It appears, then, that natural history acquired *two* paradigms in
rapid succession in the mid-eighteenth century, the Linnaean and the
Buffonian, and that these paradigms were diametrically opposed in
spirit, presuppositions, and concept of scientific method. The first,
blending Aristotelian logic and teleology with a static form of the
Christian doctrine of creation, identified natural history with taxonomy. The second, deriving from the Cartesian vision of nature as a
self-contained system of matter in motion, sought to gain insight into
this hidden system of nature by observing uniformities in the effects
it produced and constructing models capable of explaining the observed effects as necessary consequences of the operations of the
system.

The Buffonian paradigm was *not* a response to anomalies and contradictions within the Linnaean paradigm. Instead, it was a conscious
attempt to introduce into natural history concepts derived from
natural philosophy, from the seventeenth-century revolution in physics and cosmology. Buffon had had excellent training in mathematical physics and had played a significant role in the introduction
of Newtonian ideas on the Continent. In his theory of generation he
made an explicit analogy between his own organic molecules and the
atoms of Newton, between his own internal molds and the force of
gravitation.

But, despite these Newtonian analogies, Buffon was more Cartesian than Newtonian in his approach to nature. For him the uniformities observable in the motions of the planets were not, as
Newton supposed, evidences of the Creator's wise design. Instead,
they were a challenge to the natural philosopher to imagine a previous
state of the system of matter in motion from which the solar system
might have been formed by the operation of the Newtonian laws of
motion. Likewise, the uniformities which comparative anatomy discerned in the organization of animals were not to be interpreted as
evidences of design or manifestations of some transcendental idea,
but rather as occasions for constructing a theory of generation that

would display these uniformities as necessary products of the motions of organic molecules. Descartes and Leibniz, not Newton, were the sources of Buffon's paradigm of natural history.

How, then, are we to interpret the development of natural history from the death of Linnaeus (1778) and Buffon (1788) to the publication of Darwin's *Origin of Species*? Did the Linnaean paradigm predominate, develop internal contradictions and crises, and then give way to a new paradigm that had no history save the history of its development in the mind of Charles Darwin? Or did the Buffonian paradigm develop alongside the Linnaean paradigm, gradually claiming more converts until it found a decisive champion in Darwin? Or was there some interaction between the two paradigms, a thesis-antithesis relationship that eventually produced a Darwinian synthesis? Unfortunately, we know too little about the development of natural history in the nineteenth century to generalize with confidence on this subject, but it may help to consider various hypotheses in the light of what we do know.

When we examine the development of natural history in the period 1788–1859, we discover striking differences in theoretical approach to the data of natural history on both the individual and the national level. Kuhn gives us very little guidance in this kind of situation. In *The Structure of Scientific Revolutions* he treats the evolution of scientific thought and technique as if it were impervious to the influence of national cultural traditions. In another work, however, he argues that *Naturphilosophie* played a significant role in the genesis of the principle of the conservation of energy, and *Naturphilosophie* was a peculiarly German phenomenon.[2]

As will be seen, *Naturphilosophie* gave rise to something approaching a counter-paradigm in natural history in the early nineteenth century. Theories of natural selection seem to have been a purely British phenomenon in the same period. Apparently what is "normal" for the scientists of one country may be exotic from the point of view of another cultural tradition, and the cross-fertilization of ideas generated in different national contexts may play an important role in the development of scientific theory. For these reasons, as well as for the sake of convenience, we shall proceed country by country in our consideration of nineteenth-century developments in natural history.

We begin with France and the Museum of Natural History, the largest, best subsidized, best organized, best equipped establishment for the study of natural history in the world. Here, if anywhere, we should learn how a mature science develops. If we confine our attention to certain eminent figures at the Museum, notably A. L. de Jussieu and Georges Cuvier, we can make the Kuhnian model work without undue difficulty.

Building on the earlier work of Tournefort, Ray, Linnaeus, and his uncle Bernard de Jussieu, A. L. de Jussieu devised a system of botanical classification, the "natural system," that gradually gained acceptance in France, Switzerland, the Germanies, England, and America as the nineteenth century progressed. In zoology Cuvier revolutionized taxonomy by basing it squarely on comparative anatomy, Aristotelian functionalism, and Jussieu's principle of the subordination of characters. But this revolution, far from overthrowing the static paradigm of natural history, served only to strengthen and further articulate the taxonomic, teleological approach to natural history. Cuvier was proclaimed the Aristotle of the nineteenth century; his influence radiated throughout Western science.

At the same time, Cuvier dealt successfully with a major anomaly that emerged from his own researches, namely, the apparent fact that many species had become extinct. It is hard for us today to realize how anomalous this fact was for the naturalists of the eighteenth and early nineteenth centuries. In the static paradigm of natural history species had been defined as part of the stable framework of creation— "the Works created by God at first, and by him conserved to this Day in the same State and Condition in which they were first made."[3] It was inconceivable, therefore, that a species could become extinct. "For if one link in nature's chain might be lost," wrote Jefferson, "another and another might be lost, till this whole system of things should vanish piecemeal. . . ."[4]

Naturalists were extremely reluctant to envisage the possibility that species could perish, but by Cuvier's time the evidence to that effect had become overwhelming. Cuvier's work on the organic remains of the Paris basin removed the last vestige of doubt. The static paradigm of natural history was now confronted with a major anomaly demanding explanation.

It is a tribute to Cuvier's genius that he achieved a resolution of the crisis precipitated by his own researches. By extending the method and principles of comparative anatomy to the study of organic remains he simultaneously demonstrated the differences between living and fossil species and brought the latter within the domain of systematic natural history. At the same time, by adopting the geological catastrophism of Jean Deluc, he preserved the main features of the static paradigm. Species might become extinct as a result of dramatic geological upheavals of unknown origin, but in the intervals between these upheavals permanence and wise design reigned supreme, providing a stable framework for retrospective taxonomy.

By means of the doctrine of successive creations, which emerged from the researches of Cuvier, Parkinson, Buckland, and others, the static paradigm was given a new lease on life. This method of saving a paradigm by a compromise solution deserves fuller attention from Kuhn. Tycho Brahe's theory of the heavens is an earlier example of the same phenomenon.

On Kuhnian principles we should presumably look to further developments within the redefined static paradigm for the anomalies and crises that gave rise to Darwin's counter-paradigm. But the subsequent development of the "natural system" in botany and of Cuvierian ideas in zoology and paleontology failed to produce a crisis in systematic natural history. George Bentham's account of the development of botany in the first six decades of the nineteenth century is a tale of the progressive triumph of the "natural system," undisturbed by more than fleeting misgivings about the theoretical foundations of the system.[5] In zoology the chief successors of Cuvier were Owen and Agassiz, and, although both showed a tendency toward contamination by *Naturphilosophie*, neither ever doubted the essential soundness of the basic tenets of the static paradigm of natural history.

During the same half century, however, evolutionary concepts were slowly gathering momentum. From what source did they spring if not from difficulties encountered within the Linnaean-Cuvierian paradigm of *nommer, classer et décrire*? Strange to say (and this is a fact difficult to fit into the Kuhnian model), the chief source of evolutionary ideas in France during the period was the tradition of interpreting nature as a law-bound system of matter in motion, of which Buffon had been the chief exponent in the eighteenth century.

Although Cuvier established the static paradigm as the main tradition at the Museum of Natural History, the ghost of Buffon was never completely exorcised from the institution he had raised to greatness. In 1800, immediately after the death of the venerable Daubenton (to whom Buffon's ideas were anathema), Lacépède and Lamarck both published evolutionary speculations similar in outlook to those of their mentor Buffon. Twenty-five years later Étienne Geoffroy St. Hilaire, having already fallen afoul of Cuvier by his advocacy of transcendental anatomy, turned to evolutionary speculations of a distinctly Buffonian character. Indeed, one of his last publications contained an appreciation of Buffon.[6]

Concerning Lacépède's evolutionism little need be said, since Lacépède was soon drawn away from natural history into the Napoleonic administration, where, unlike Cuvier, he found little time for scientific research.[7]

Lamarck, on the contrary, was a major figure in the development of scientific natural history. Trained by Bernard de Jussieu, befriended by Buffon (whose son he tutored), Lamarck made a sufficient reputation as a botanist to be appointed to the chair of invertebrate zoology at the Museum of Natural History. There, less than ten years after his appointment, he sketched the outlines of a general "physics of the earth," embracing "all the primary considerations of the earth's atmosphere, of the characteristics and continual changes of the earth's external crust, and finally of the origin and development of living organisms." The first part of this science he called Meteorology, the second Hydrogeology, and the third Biology. Biology, he explained, was not to concern itself primarily with taxonomy, but rather with discovering the causes, laws, and direction of organic change.

How are we to regard Lamarck's effort to redefine the basic concepts and goals of natural history? From a Kuhnian point of view this was certainly an attempt at a scientific revolution affecting every level of natural history from the concept of a species to the definition of the ultimate goals of the naturalist. It had all the characteristics of a scientific revolution except success.

Unfortunately, Kuhn's analysis lays down no guidelines for dealing with unsuccessful revolutions in science. We cannot deny that Lamarck's ideas were revolutionary. In broad outline—geological uniformitarianism with its vast time scheme, descent with modification by natural causes, progressive development up to and including man, the search for the laws and causes of organic change—they were similar to the ideas Darwin was to champion half a century later. But the proposed mechanism of organic change was unconvincing, and the circumstantial evidence supporting the theory was scanty and sporadic.

Yet one feels entitled to ask why this revolution should have been attempted at this time. Was there an anomaly-generated crisis in natural history in the late eighteenth century that gave rise to Lamarck's counter-paradigm and the less fully articulated counter-paradigms of Erasmus Darwin, Lacépède, and others? It is hard to believe that any such crisis existed at that time. Evidence indicating widespread extinction of species did not become available until *after* Lamarck had arrived at the grand outlines of his theory, and, in any case, Lamarck did not believe that species became extinct. (So-called extinct species were for him simply the ancestors of living forms.)

Likewise it seems unlikely that some anomaly or crisis in botanical or zoological taxonomy drove Lamarck to an evolutionary position, although Charles Lyell was later to ascribe Lamarck's evolutionism to the difficulties he encountered in distinguishing species from varieties.

On the whole, it seems more likely that Lamarck became convinced of the mutability of organic forms from his geological researches, and that he guessed the direction of organic change partly from the old idea of the scale of nature and partly from the researches of Jussieu and Cuvier on the anatomy of plants and animals.

Like Buffon, Lamarck started from the idea that nature was a lawbound system of matter in motion. But, whereas Buffon attributed the main features of the earth's surface to the action of tidal currents operating on the plastic surface of a cooling globe, Lamarck invoked the action of running water and the progressive displacement of ocean basins by the action of waves. As a result, he adopted a time scheme of millions of years for earth history in place of the tens of thousands envisaged by Buffon. Nature, he declared, had plenty of time and circumstances at her disposal, and everything conspired to prove that all of her works, even the largest and most imposing, were subject to slow change.

Thus, Lamarck drew from a thoroughgoing geological uniformitarianism the inevitable conclusion that organic forms were mutable. In the static paradigm of natural history inorganic, organic, and human history were assumed to be synchronous. The basic structures of inorganic nature were thought to have been perfectly contrived by the Creator to subserve the needs of the higher levels of existence, animate and rational. But if Lamarck and Hutton were right, if the inorganic environment had been undergoing constant slow change for millions and millions of years as a result of geological processes governed by the general laws of physics and chemistry, with no vestige of a beginning, no prospect of an end, then the organisms inhabiting the globe must have changed too, or they would have suffered extinction.

Thus the first serious crisis in the static paradigm of natural history arose, not within the system of naming, classifying, and describing which constituted the heart and soul of the paradigm, but rather from a postulate affecting the wider framework of assumptions concerning the stability of the visible structures of nature and their hierarchical ordering with respect to each other. Being geologists rather than naturalists, Hutton and Playfair did not develop the implications of geological uniformitarianism for systematic natural history. But Lamarck could not escape them. Either he must accept the wholesale extinction of species as a logical consequence of his geological ideas, or he must conceive the organisms inhabiting the earth's surface as being endowed with a natural capacity to undergo the changes required for survival amid changing circumstances and seek to discover the means by which these changes had been effected.

Of these two alternatives Lamarck chose the latter, invoking use and disuse as the chief agencies of organic change. His next task was to explore the implications of his revolutionary hypothesis for taxonomy and the idea of a natural method of classification. Here Lamarck seems to have taken his clue from the old idea of the scale of nature. In any case, he devoted his energies chiefly to working out a classification that would reflect the path nature had followed in giving birth to progressively more complicated living forms.

It appears, then, that Lamarck's counter-paradigm sprang more from a predisposition toward a uniformitarian view of nature's operations than from a sense of difficulties to be resolved in the structure of systematic natural history. This is not to say that Lamarck did not find the evolutionary postulate useful in taxonomy, but only that taxonomic problems were probably not the main source of his belief in the mutability of species.[8]

However that may be, the question remains as to what role, if any, Lamarck's theory played in the eventual emergence of the Darwinian counter-paradigm. It is fashionable nowadays to deny Lamarck any status as a precursor of Darwin, but we had best postpone this question of the influence of Lamarck's *révolution manqué* on Darwin's *révolution véritable* until we deal with developments in Britain. Suffice it to say for the moment that Lamarck's ideas haunted natural history during the first half of the nineteenth century much as the spectre of communism haunted social theory in the second half. Darwin himself once referred to Lamarck as "the Hutton of geology," obviously intending to write "the Hutton of biology." That was a high compliment and a shrewd characterization. In extending the uniformitarian concept from geology to biology Lamarck foreshadowed the doom of the static paradigm of natural history.

Before turning to developments in Britain that proved decisive for the overthrow of the static paradigm, we must consider briefly another attempt at revolutionizing the conceptual framework of natural history, namely, the attempt associated with the rise of German *Naturphilosophie*. Although Goethe, Oken, Carus, and their followers did not break with the main tradition of systematic natural history as sharply as Lamarck did, their deviation from some of its basic tenets approached the dimensions of a genuine counter-paradigm and gave rise to a kind of evolutionism.

Naturphilosophie diverged from the Linnaean-Cuvierian tradition in natural history in several important respects.

1. It rejected the teleological functionalism of the dominant tradition in favor of a science of pure form, in which form was conceived as dictating function, not function form. The Cuvierian techniques in comparative anatomy were retained, but they were employed for a new purpose: the discovery of a uniform plan of organization pervading the organic world. Instead of the correlation of parts and their adaptation to each other and to the conditions of existence, the watchwords of transcendental anatomy were unity of plan and the correspondence of parts. Classification was still important, but it was subordinated to the search for archetypes.

2. Emphasis on development, on nature begetting, supplanted the traditional preoccupation with the description and classification of begotten forms. On the whole, the idea of development was restricted to embryological development, but embryological study led on to the idea of parallelisms between the levels of organization traversed in embryological development and the levels of organization revealed by comparative anatomy and (eventually) by the fossil record. By some writers these parallelisms were given an evolutionary interpretation.

3. Creationism was muted or abandoned outright in favor of pantheistic ideas of creative nature, spontaneous generation, and the like. Whereas in England most naturalists considered belief in spontaneous generation unscientific and atheistical, many German writers considered the doctrine of successive creations unscientific, preferring to resort to successive spontaneous generations to explain the changes in flora and fauna revealed in the fossil record.

How are we to regard these developments in the light of Kuhn's model? Was this another abortive revolution provoked by anomalous discoveries in systematic natural history? Such an interpretation would be hard to sustain. *Naturphilosophie* was an outgrowth of German idealistic philosophy. Perhaps it developed in reaction to certain aspects of the thought of the Enlightenment, as Charles Gillispie has suggested. But do we solve the problem of how science develops by casting whole scientific movements into the outer darkness with the label "subjective science" attached to them?

Naturphilosophie enlisted many able scientists under its banner. Its influence was felt throughout Europe, even in England, where

Richard Owen became its leading exponent. In France, Étienne Geoffroy St. Hilaire elaborated a science of pure form independently of the German writers, and Geoffroy can scarcely be described as anti-Enlightenment. Shall we not rather say that the idea of an all-embracing unity of plan in the organic world was a natural and legitimate product of scientific imagination seeking ever wider generality in its ordering of nature?

But here again, as in the case of the very different visions of nature and natural science promulgated by Buffon and Lamarck, paradigm construction did not wait on the emergence of anomalies and crises in systematic natural history. On the contrary, it ran ahead of known facts, postulating a wider unity in nature than could be demonstrated and delving into the study of embryological development in search of confirmatory data.

In the end, the devotees of transcendental anatomy failed to establish their case. But their researches and many of their concepts, such as the ideas of homology, recapitulation, balancement, and even evolution, entered into the general fund of knowledge and speculation available to Darwin and his contemporaries.

We come now to Britain, where the main revolution in natural history was to take place, although from an unexpected quarter. Systematic natural history made slow progress in Britain after the brilliant work of John Ray in the late seventeenth century. Linnaean influence did not become entrenched there until the 1780's, when Sir James Edward Smith acquired a vested interest in Linnaean botany through his purchase of Linnaeus' herbarium, books, and letters and joined with the Reverend Samuel Goodenough and others in founding the Linnaean Society of London. In zoology, George Shaw adopted the Linnaean classification in preference to Thomas Pennant's system based on Ray.

As the nineteenth century wore on, British science responded to developments on the Continent and began to make solid contributions to the literature of natural history. In botany the so-called natural system of Jussieu and Candolle was gradually introduced by Robert Brown, John Lindley, George Bentham, and the Hookers, and Kew Gardens began to emerge as a center for the study of world botany. In zoology the dominant figure was Richard Owen, who combined Cuvier's techniques in comparative anatomy with the transcendental ideas of Oken, Goethe, and Geoffroy St. Hilaire in a way that would have dismayed Cuvier.

There was nothing very revolutionary in all this, nor does one sense a spirit of unrest or crisis among these naturalists. In Bentham's eyes the period before Darwin was characterized by the progressive triumph of the "natural system," leading many botanists to think that little was left for systematic botany but mopping up operations. Joseph Dalton Hooker was privy to Darwin's subversive hypothesis from 1844 on, but Darwin's powerful ideas worked slowly on Hooker's imagination. Apart from his intercourse with Darwin, Hooker would never have broken out of the static paradigm of natural history. In zoology there was equally little evidence of a crisis psychology. The *Transactions* of the Zoological Society of London and the periodic progress reports of the British Association for the Advancement of Science gave little hint of the coming revolution. True, there was an extraordinary outcry against *The Vestiges of the Natural History of Creation*, but Chambers' book was a challenge from outside the natural history establishment, not from within.[9]

What, then, were the sources of the British revolution in natural history, if they are not to be found in the internal development of systematic botany and zoology? One might be tempted to find them in the eighteenth-century tradition of speculative philosophy of nature represented by Erasmus Darwin, but Erasmus Darwin's scientific impact, as distinguished from his popular influence, was negligible. Of far greater consequence for the evolution of natural history were certain developments in British geology and political economy in the years from 1775 to 1835.

Let us speak first of the progress of geology. The immediate impact of Hutton's uniformitarianism on systematic natural history was minimal, partly because his ideas were not widely accepted, but also because neither Hutton nor Playfair made more than passing reference to the organic remains embedded in the crust of the earth. Nevertheless, as we have seen in our discussion of Lamarck, geological uniformitarianism had momentous implications for the doctrine of the fixity of species. These implications were not lost on Playfair, as can be seen in the following passage from his *Illustrations of the Huttonian Theory of the Earth*, a passage used by Lyell as a motto for the second volume of his *Principles*:

> The inhabitants of the globe, then, like all the other parts of it, are subject to change: It is not only the individual that perishes, but whole *species*, and even perhaps *genera*, are extinguished. . . . But besides this, a change in the animal kingdom seems to be a part of the order of nature, and is visible in instances to which human power cannot have extended.[10]

For the time being, however, British naturalists and paleontologists evaded the issues posed by geological uniformitarianism in the same way that Cuvier evaded them in France. They rejected uniformitarianism in favor of a theory of successive creations and extinctions and devoted themselves to naming, classifying, and describing the organic remains of former worlds and to discovering how to identify and correlate geological formations by means of them. In so doing they unwittingly set the stage for Charles Lyell's revolutionary extension of the uniformitarian doctrine to organic phenomena.

There is little evidence, however, that Lyell's great book was a response to a state of crisis in geological science. Instead, it seems to have been conceived as a reaffirmation of uniformitarian principles, and, what was crucial for the development of evolutionary ideas, an extension of them to the organic world, at least in regard to the extinction of species. Lyell's shift to a dynamic and causal view of organic nature is apparent in the opening paragraph of his *Principles*, where he says: "Geology is the science which investigates the successive changes that have taken place in the organic and inorganic kingdoms of nature; it inquires into the causes of these changes, and the influence which they have exerted in modifying the surface and external structure of our planet. By these researches . . . we acquire a more perfect knowledge of its present condition, and more comprehensive views concerning the laws now governing its animate and inanimate productions."[11] Lamarck himself could not have asked for a better statement of the aims and outlook of a comprehensive science of the earth and its productions.

But Lyell was not prepared to follow Lamarck down the uniformitarian path to a full-blown evolutionism. Like Lamarck, he drew from geological uniformitarianism the conclusion that plant and animal species must change with changing circumstances or perish. But, whereas Lamarck viewed organisms as endowed with an innate capacity to undergo the changes necessary for survival, Lyell, unconvinced that organisms possessed an unlimited capacity for variation, chose instead to envisage piecemeal extinction of species as the eventual consequence of their limited ability to adapt to changed conditions. Thus, whereas Lamarck's energies were directed toward imagining the processes by which organisms adapted to changing circumstances and toward tracing the path of their upward evolution, Lyell's were concentrated on studying the effects of environmental changes on the chances of survival of species possessing limited powers of variation. Not evolution, but elimination in the struggle for survival, became the focus of his attention so far as species were concerned.

This was precisely the direction of thought that was to eventuate in the theory of natural selection. Moreover, it was a mode of thinking that came naturally to Englishmen, steeped as they were in the tradition of Adam Smith, Malthus, and Ricardo. Surely it is no mere coincidence that all of the men who arrived at some idea of natural selection in the first half of the nineteenth century—one thinks of William Wells, Patrick Matthew, Charles Lyell, Edward Blyth, Charles Darwin, A. R. Wallace, and Herbert Spencer—were British. Here, if anywhere in the history of science, we have a striking example of the influence of national habits of thought on the development of scientific theory, a phenomenon difficult to reconcile with Kuhn's internalist approach. For the cast of mind we have been describing affected not merely the timing of the revolution in natural history but its central concept, the idea of competition, survival of the fittest, and consequent progress.

Lyell himself stopped short of a theory of the origin of species, falling back on the traditional belief in special creation and wise design. But his uniformitarian explanation of the piecemeal extinction of species seemed to cry out for a correlative explanation of the origin of species by natural causes. The elements of a non-Lamarckian theory of evolution, stressing the struggle for existence and survival of the fittest, were present in his work cheek and jowl with his systematic exposition and discussion of the Lamarckian alternative to his own steady-state concept of earth history. Little wonder, then, that Lyell's *Principles* provided the impetus for evolutionary speculation in Britain from the time of its publication onward.

Chambers and Spencer, impressed more by Lyell's exposition of Lamarckian ideas than by his refutation of them, chose Lamarck's kind of evolutionism. Darwin and Wallace, aware of the inadequacy of Lamarck's theory of organic change but convinced of the essential truth of transmutationism, set out to discover the mechanism of change. Meanwhile, systematic natural history continued on its accustomed course, untroubled by any sense of crisis. A revolution was impending, but it was to come from outside, not from within, the establishment.

We come now to Darwin and the revolution in natural history associated with his name and achievements. From Kuhn's hypothesis we should expect this revolution to be non-cumulative in character and to involve the substitution of a new paradigm of natural history incommensurable with the static paradigm that had reigned before the revolution. The first question, then, is: Was the Darwinian revolution non-cumulative in character? That is, did it break sharply with the concepts, methods, and modes of thought that had prevailed before 1859?

From what has already been said it should be apparent that this question does not admit of a simple Yes or No answer. If we compare Darwin's ideas and methods with those that had prevailed in the main tradition of systematic natural history, namely, the tradition of Linnaeus, Jussieu, Candolle, Cuvier, Owen, and Agassiz, we discover a profound break with the past, though *not* one generated in response to internal difficulties in the tradition that was overthrown.

If, on the other hand, we compare Darwin's concepts and methods with those of Buffon, Erasmus Darwin, Lamarck, Étienne Geoffroy St. Hilaire, and Charles Lyell, we begin to have doubts about the non-cumulative character of the Darwinian revolution. We discover that some aspects of Darwin's thought and practice were more original than others.

Darwin himself made no claim to have invented the idea of organic evolution. He was too well acquainted with the writings of Lamarck, Geoffroy St. Hilaire, and his own grandfather, Erasmus Darwin, all of whom he had read or re-read on his return from the voyage of the *Beagle*, to make any such claim. He claimed only to have discovered "the means of modification and co-adaptation" in nature and thereby to have transformed a speculative idea of descent with modification into a workable theory of the origin of species. To this he might have added that he had done more than merely hit upon the idea of natural selection as the means of modification and co-adaptation. More important, he had deduced the consequences of his hypothesis and endeavored to show by observation and experiment that they actually obtained in nature. This combination of inductive and deductive methods had long prevailed in the physical sciences, but Darwin was the first to apply it systematically in natural history. His methods were as revolutionary as his theory.

Thus, one's judgment as to the cumulative or non-cumulative character of the Darwinian revolution depends largely on whether one stresses the general concept of descent with modification or the particular theory of natural selection as the means of organic modification in nature. The general concept had a history reaching back at least to Buffon. Indeed, Thomas Henry Huxley was inclined to credit Descartes with insight into what Huxley deemed "the fundamental proposition of Evolution," namely, that "the whole world, living and not living, is the result of the mutual interaction, according to definite laws, of the forces possessed by the molecules of which the primitive nebulosity of the universe was composed."[12] In one sense Huxley was right. Geological uniformitarianism and its corollary of indefinite mutability in the organic world were implied in the Cartesian program of deriving the present structures of nature from a simpler, more homogeneous state of the system of matter in motion by the operation of the laws of nature. The drawing out of this implication by Buffon, Lamarck, Lyell, and others sprang more from the appeal of this vision of nature and natural science to imaginative minds than it did from factual discoveries, which could always be interpreted differently by less imaginative observers.

Darwin himself conceded the importance of this dynamic and causal approach to nature when he wrote in 1863: "Whether the naturalist believes in the views given by Lamarck, by Geoffroy St. Hilaire, by the author of the 'Vestiges,' by Mr. Wallace or by myself, signifies extremely little in comparison with the admission that species have descended from other species, and have not been created immutable: for he who admits this as a great truth has a wide field opened to him for further inquiry."[13]

The difficulty was, however, that there could be no general acceptance of the idea that species have descended from species until someone could show convincingly *how* this could take place. The revolution in natural history had been prophesied for more than a century, but the fulfillment of the prophecy had to wait on the discovery and elaboration of a theory of natural selection. It was Darwin and Wallace who achieved this result, but, as we have seen, the way was cleared for them by Charles Lyell and by the British school of political economy. Patrick Mathew, Edward Blyth, and Herbert Spencer were products of the same climate of opinion.

Yet, curiously enough, although the theory of natural selection played an indispensable role in converting the scientific world to an evolutionary point of view, many who accepted transmutationism after Darwin rejected natural selection as the key to organic evolution. Neo-Lamarckian evolutionists, some of them distinguished scientists, were numerous in the late nineteenth century.

This fact brings us to the final question, whether Darwin's work effectively established a new paradigm incommensurable with the static paradigm of systematic natural history. If by establishing a new paradigm we mean simply establishing an evolutionary point of view in biology, Darwin certainly did that, and the new point of view can justly be described as incommensurable with the Linnaean paradigm, although no more incommensurable with it than the point of view of Buffon or Lamarck was. If, on the other hand, we mean that Darwin's work established the theory of natural selection and Darwin's general

assumptions and methods as the norm of scientific thought and practice among biologists, this is a more dubious proposition.

As we have seen, many scientists accepted Darwin's evolutionism
but not his emphasis on natural selection. Among systematists, moreover, although lip service was now paid to the Lamarckian and Darwinian idea that the natural method of classification was one that
reflected phylogeny, taxonomic methods were slow to change. The
anthropologists, far from following Darwin's lead in investigating the
origin of human races, settled down to three-quarters of a century of
hairsplitting racial taxonomy. As for Darwin's ideas on the subject
of heredity and variation, they did give rise to a school of English
geneticists led by Galton, Pearson, and Weldon. But, far from being
generally accepted, these ideas were attacked strenuously by Weis-
mann and others and were eventually overthrown by the rediscovery
of Mendel's laws and the development of cytology. In fact, it could
be argued that nothing approaching a "Darwinian" paradigm became established until the 1930's, and even that paradigm was Darwinian only in a very loose sense.

We conclude as follows:

1. Through the work of Ray, Tournefort, and Linnaeus natural
history acquired a conceptual framework that dominated the study
of natural history until Darwin published his *Origin of Species*, after
which systematics was gradually reshaped and relocated within the
broader framework of evolutionary biology.

2. Challenges to the dominance of the Linnaean framework in
natural history arose both within and outside of that framework.

3. The challenges arising within the static view of nature and
natural history failed to precipitate a search for new premises; they
were either ignored or evaded by compromises such as the theory of
successive creations.

4. Rival concepts, such as those propounded by Buffon, Lamarck, and the transcendental anatomists, arose from time to time,
but not in response to anomalies or crises within the dominant view.
These counter-concepts exerted a significant influence both on the
static view of nature and on the developments that were to eventuate
in its overthrow.

5. The evolutionary alternative to the static outlook developed
chiefly outside the Linnaean framework in the form of a search for a
science of nature that would derive the phenomena of nature from
the operations of a law-bound system of matter in motion. The earliest and most powerful challenge to the static view of nature was the

challenge implied in geological uniformitarianism. It was this postulate, rather than particular scientific discoveries, that drove Lamarck
to an evolutionary position and led Lyell to envisage the piecemeal
extinction of species through the struggle for existence.

6. The eventual emergence of the theory of natural selection in
Britain seems to have owed a great deal to the influence of the competitive ethos that pervaded British political economy and British
mores generally.

7. The Darwinian revolution displayed elements both of continuity and discontinuity with the past. It overthrew the static view of
nature and natural history, but failed to establish a clear-cut paradigm
in its place.

8. The Kuhnian paradigm of paradigms can be made to fit
certain aspects of the development of natural history from Ray to
Darwin, but its adequacy as a conceptual model for that development
seems doubtful. The use of Kuhnian terminology in this essay should
not be interpreted as implying belief in its general utility for the
historiography of science. At the same time, it should be remembered
that an inadequate hypothesis is better than none at all. Those who
question the validity of Kuhn's model should feel themselves challenged to provide alternative interpretations of the genesis of revolutions in science. The present essay is intended less as a critique of
Kuhn's stimulating book than as a tentative formulation of some
general ideas about the rise and development of concepts of organic
evolution.

1. Carl Linnaeus, *A General System of Nature . . . Translated from
Gmelin's Last Edition . . .* , William Turton, tr. and ed. (London, 1806),
I, 2; Georges Louis Leclerc, Comte de Buffon, *A Natural History, General and
Particular . . .* , William Smellie, tr. (3rd ed., London, 1791), VIII, 33–34.

2. Thomas S. Kuhn, "Energy Conservation as an Example of Simultaneous Discovery," in Marshall Clagett, ed., *Critical Problems in the History of Science* (Madison, Wisconsin, 1959), 321–56.

3. John Ray, *The Wisdom of God Manifested in the Works of the Creation. . . .* (3rd ed., London, 1907), Preface (not paged).

4. Thomas Jefferson, "A Memoir on the Discovery of Certain Bones of
a Quadruped of the Clawed Kind in the Western Parts of Virginia," *Trans.
Amer. Philos. Soc.*, IV (1799), 255–56.

5. George Bentham, "On the Recent Progress and Present State of Systematic Botany," *Report of the 44th Meeting of the British Association for
the Advancement of Science in 1874* (London, 1875), 31 ff.

6. Étienne Geoffroy St. Hilaire, *Fragments biographies précédés
d'études sur la vie, les ouvrages et la doctrine de BUFFON* (Paris, 1838),
3–157. In this essay Geoffroy St. Hilaire exhibits in a striking manner the con-

trast between the Linnaean-Cuvierian concept of natural history and that of Buffon and presents Lamarck's evolutionary philosophy and his own belief in the mutability of species as a continuation of Buffon's vision of nature as a system of *faits nécessaires*.

7. Bernard G. E. de la Ville sur Illon, Comte de Lacépède, *Histoire Naturelle des Poissons* (Paris, 1798–1804), II, 9–68.

8. The mental process by which Lamarck arrived at an evolutionary viewpoint in the last two or three years of the eighteenth century is difficult, if not impossible, to ascertain, but Lamarck's own statements lend considerable support to the interpretation I have advanced. In the opening sentences of the Appendix to his *Discours D'Ouverture de L'An X* he says:

> I thought for a long time that there were constant species in nature, and that they were constituted of the individuals belonging to each of them.
>
> Now I am convinced that I was in error in this regard, and that there are really only individuals in nature.
>
> The origin of this error, which I shared with many naturalists who still hold to it, is found in the *long duration* with respect to us of the same state of things in each place which each living body inhabits; but that duration of the same state of things for each place has a limit, and with plenty of time it undergoes mutations at each point of the surface of the globe which change the circumstances for all the living bodies inhabiting it. . . . Elevated places are constantly degraded, and everything which is detached is carried toward the low places. The beds of rivers, of streams, of seas even are insensibly displaced, as well as climates; in a word, everything on the surface of the earth changes little by little in situation, in form, in nature and aspects. . . . [Here he cites his *Hydrogéologie*]

Thus, Lamarck argues that the mutability of species follows from the mutability of the earth's surface and that both types of mutability are hidden from man by the extreme slowness of the changes that take place. In his *Discours D'Ouverture de L'An XI* (pp. 541–42) he makes it clear that organisms are *forced* to undergo change as a result of changes in their environments: " . . . we know . . . that a *forced and sustained change*, whether in the habits and manner of living of animals, or in the situation, the soil and the climate of plants, effects after a sufficient time a very remarkable mutation in the individuals exposed to it." (Italics mine.)

Finally, in his essay "Sur Les Fossiles" in his *Système Des Animaux Sans Vertèbres*, Lamarck reveals his awareness that, given the changes on the earth's surface postulated by geology, the alternative to transmutation of living forms is widespread extinction of species. Some naturalists, he says, have concluded from the lack of perfect resemblance between fossil and living species "that this globe has undergone a universal *bouleversement*, a general catastrophe, and that as a result a multitude of species of animals and of plants have been absolutely lost or destroyed." But Lamarck will have nothing to do with such a universal catastrophe, "which by its very nature regularizes nothing, confounds and disperses everything, and constitutes a very convenient means for naturalists who wish to explain everything and who do not take the trouble to observe and study the process which nature follows in regard to her productions and in everything which constitutes her domain." Instead, he undertakes to show "that although many of the fossil shells are different from all known marine shells, this by no means proves that the species of shells have been obliterated, but only that these species have changed in the course of time." ["Sur Les Fossiles," *Système Des Animaux Sans Vertèbres*, (Paris, 1801), 408–409. Translation mine.]

Thus, according to Lamarck, geology and paleontology present the naturalist with a choice: either (1) catastrophism and wholesale extinction of species, or (2) transmutationism on uniformitarian principles with little or no extinction of species. His own decision favors the second alternative, not only because of his uncompromising uniformitarianism in every branch of natural history but also because he finds in the transmutation hypothesis a neat solution to grave problems in taxonomy, an exhilarating sense of progress in nature, and a vindication of the wisdom of the Author of Nature, a wisdom which wholesale extinction of species would seem to impugn.

He seems never to have considered the possibility of combining uniformitarianism with acceptance of widespread extinction of species. It would take an Englishman to see the wise dispensation of the Creator in the competitive struggle for existence. [The passages from Lamarck's inaugural discourses are translated from the *Bulletin Scientifique de la France et de la Belgique*, XL (1906).]

9. This is not to say that there was no undermining of belief in the fixity of species among naturalists as the nineteenth century progressed. In botany Schleiden, Unger, and Rafinesque accepted transmutation by mid-century. But Darwin worked out his theory of natural selection in the thirties, largely in response to problems he had encountered on the voyage of the *Beagle*. Unfortunately, there has been too little research on the period 1830–1859 to warrant confident generalizations about the state of mind of naturalists in this period. My impression is that it would be difficult to prove the existence of a "crisis psychology" in the 1850's and impossible to do so for the 1830's.

10. John Playfair, *Illustrations of the Huttonian Theory of the Earth* (Edinburgh, 1802), 469–70. I have used the facsimile reprint of this work (Dover Publications, Inc., New York, 1964). Playfair envisages the modification as well as the extinction of species driven into new habitats by man: " . . . the more innocent species fled to a distance from man, and being forced to retire into the most inaccessible parts, where their food was scanty, and their migration checked, they may have degenerated from the size and strength of their ancestors, and some species may have been entirely extinguished."

11. Charles Lyell, *Principles of Geology* . . . (2nd ed., London, 1832), I, 1.

12. Thomas Henry Huxley, "Evolution in Biology," *Darwiniana: Essays* (New York, 1908), 206. This essay was originally published in 1878.

13. Letter from Charles Darwin to the *Athenaeum*, Down, England, May 5, 1863, quoted in Francis Darwin, ed., *The Life and Letters of Charles Darwin* . . . (New York, 1898), II, 207.

Commentary on the Paper of John C. Greene

By William Coleman, The Johns Hopkins University

PROFESSOR GREENE has adopted the "Kuhnian paradigm," a murky entity we must concede at the outset, and explores its applicability to the historical development of an equally ill-defined subject, natural history. His exposition of Kuhn's conception of normal science and scientific revolution is ample and generally fair. The tone of Professor Greene's statement is cordial and seemingly laudatory. His objective, however, lies elsewhere. Professor Greene will show that the representation of scientific change by "paradigm shifts," however beguiling it be on the abstract level, is too loosely defined and too poorly exemplified to offer meaningful substance to the historian of the natural sciences.

The good historian deals with the "complexities of the concrete historical process." This is Professor Greene's expression and doubtlessly few of us will disagree with it. Its implications, which Professor Greene has had opportunity only briefly to expound, are a severe trial to the paradigm hypothesis, especially if advanced in rigid form. Now Professor Greene suggests that Kuhn's model "makes no provision" for the simultaneous activity of numerous and probably opposing paradigms. This situation offers, of course, the very stuff of historical interest and inquiry; it virtually *is* that complexity into which historians, myself included, so like to submerge. Kuhn has not, I believe, so clearly excluded this possibility. He does not emphasize "monolithic" science. Every paradigm is loosely articulated; conflicting paradigms can and do coexist. Science, he tells us—and this is really one of his most important points—is rather a "ramshackle affair."

Here we are at the heart of much criticism of Kuhn's hypothesis. To be interesting we would hope, we trust, that any prosperous period of normal science depends on one or at most very few paradigms. It is quite fair to claim that normal science simply demands at least this much. The historian, then, looks to particular concrete situations. He finds a paradigm. But no! He finds two paradigms, three, four, and so on. Thus Professor Greene has identified as paradigmatic a Static, a Dynamic, a Naturphilosophische and, later, a competitive conception of organic nature. There are still further alternatives; I shall later note one of these. The suggestion here is clear: once we press on beyond an imputed and probably unitary world-view characteristic of an epoch we begin to encounter diversity of opinion. That diversity becomes increasingly radical, so much so that we are incapable of re-erecting from it our original and seemingly global view—save by abuse of logic and historical context and through consummate faith in our initial structures.

Kuhn seems aware of the problem, but perhaps not so acutely as his critics, including Professor Greene. Gerd Buchdahl has returned the issue to a familiar one regarding theories. If the paradigm is elevated to a regulative role, that is, it is, and sets, the order-of-the-day for scientific activity, it becomes virtually *all* scientific thought. If the paradigm is deemed more specific, then it deals with particular scientific problems, accounting for data, perhaps predicting new phenomena and doubtlessly appealing to the scientist by its simplicity and "fitness." The latter situation, claims Buchdahl, is that usually discussed about the term theory. Kuhn, for many and, in my opinion, necessary reasons, wants to escape that word theory. It smells too much of unwholesome and to date excessively influential brands of philosophies of science. It distorts our conception of what science is, does, and has been.

But Buchdahl's charge remains and we can see in Professor Greene's account a reaffirmation of the point. Again, if the paradigm is to be interesting it must indeed prevail, prevail overwhelmingly in a period of normal science. It must be general and is by definition regulative.

We have already heard Professor Greene's roster of viable and opposing paradigms for the early period of natural history. Kuhn has two routes of escape. The first is to argue that the seventeenth and eighteenth centuries are pre-paradigmatic, hence unsystematized and groping, for natural history. I doubt that Kuhn would claim any such thing, and Professor Greene would cry foul if it were done. The other option is to relax one's definition—the implicit one—of paradigm. As Kuhn suggests, there may be greater or lesser paradigms in currency during a given period. Here is an opening—it allows us to apply paradigm language to at least the more important and consistent vagaries of thought of a given period. But the price is high, and this is why I believe any good historical analysis will find the whole paradigm conception short on meaningful substance.

I shall put it bluntly. The importance of the paradigm lies in its relationship to ongoing or incipient normal science. Normal science is essentially a body of belief and practice of a scientific community. The notion of community is all-important to Kuhn, derives from other realms of social and linguistic philosophy and presents us with the idea of structures. Communities, therefore, entertain paradigms and, conversely, paradigm exploitation is the business of the community. The same community should have but one paradigm, and the contemporaneous existence of several paradigms suggests so many other communities. The old theme returns: historical situations are complex, paradigms and communities many and reliable historical investigations demonstrably few.

Using paradigms on the general level we only repeat interesting truisms: the paradigm shift in the Copernican revolution, for example, represents a radical alteration of viewpoint—from geocentric to heliocentric—from which to regard planetary phenomena. On the specific level our paradigms multiply; we have, for example, revolutions in chemistry based on a shift from phlogiston to oxygen or on Dalton's atomic hypothesis. We think we have said something more profound when these familiar events, and hosts of others less obvious, can be brought together by the paradigm-normal science hypothesis. I for one believe we have gained very little, save the indefensible assertion that so many and such diverse "paradigms" are indeed commensurable.

But this is really Professor Greene's conclusion and it is, I may be so reckless to suggest, inherent in Kuhn's over-all argument. Both seem in agreement, and Kuhn here is more than explicit, that the enemy is the Whig interpretation of history, the celebrated and ever-prosperous idea that history's meaning is fully betrayed by its relation to the present (or future). As Kuhn remarks, we have too often considered science as developing *toward* particular goals rather than treating thoroughly and with whatever dispassion we might muster the process of change as science *moves from* one stage or another. Confirmation theory and the falsifiability doctrine, the strong points in the philosophers' fortress, Kuhn shows as supporting this unjustifiable view of science as an incremental process. Each new small truth won and each falsehood destroyed do not carry us progressively toward Truth itself; they are only components, and important components, of the daily round of normal science.

We are thus returned to a world of historical relativism. Standards external to time, place, society and contemporary thought offer no grounds for useful judgment. Science again becomes the activity of human beings, trained by and finding their place in the guild but nonetheless, and happily, prey to a vast spectrum of conflicting motives and patterns of thought. This is to decide—willfully and not rigorously!—the ancient debate over Universals and Particulars in favor of the latter.

Kuhn, I'm sure, would be horrified by such rank subjectivism; Professor Greene may also choose to flee this conclusion. In any case, the paradigm hypothesis when pushed hard either lifts us up and betrays the obvious, or brings us down to the tangible realm, if such it be, of historical time and place. This may not be the role envisaged for history of science by its many friends, but it may promote our discipline's well-being and importance in other desirable ways.

I now turn abruptly and more briefly to Professor Greene's roster of natural history paradigms, retaining that term for convenience's sake alone. He has identified first a static paradigm. Advocated by Linnaeus, it espied order in nature and ascribed it to divine wisdom and power. The naturalist's task was to name, classify and describe the seeming diversity of minerals, plants and animals.

The static paradigm arose in the seventeenth, prospered in the eighteenth and continued well into the nineteenth century. But from mid-eighteenth century it faced a formidable rival. Buffon proclaimed a dynamic view of nature. In Professor Greene's words, Buffon felt the main business of natural history was "to explain the observed uniformities in nature's productions as necessary results of the operations of the hidden system of laws, elements and forces" of nature. The Buffonian wanted causal explanation, not simple categorization of changeless entities.

Circa 1800, a third option enters; it has the "dimensions of a genuine counter-paradigm." This is the paradigm of the German nature-philosophers. They stressed, amongst unmentionable metaphysical audacities, a science of pure form, explanation in terms of development and pantheistic ideas of creative nature.

Finally, hints of a fourth paradigm intrude—that of an English nation intent on markets and profits and therefore keenly concerned with the competitive aspects of human society and organic nature.

Professor Greene argues that these paradigms are either contemporaneous or have not, as the Kuhnian model urges, yielded to one another via a process of paradigm articulation, anomaly, crisis and revolution or true innovation, the last event remaining, inci-

dentally, "inscrutable." Professor Greene's tacit acceptance, and subsequent disapproval, of that model, together with earlier remarks,
persuades me that such activity is premature. We have to buy the
possibility and utility of the framework before building our house,
and that possibility and utility remain much in doubt.

But what should we do with these ostensible paradigms for natural
history? I suggest that we severely question them, not for their designation as paradigms but for the content which Professor Greene has
assigned them. The Buffonian paradigm, in particular, seems much
too inclusive or, put differently, if it is narrowly defined much important eighteenth-century thought is excluded—and it will find no
place in the static paradigm either. The core of the Buffonian paradigm was the Cartesian notion of a "lawbound system of matter and
motion." But matter and motion are only one way of accounting for
change in nature. Both require further specification. Both may indeed
be but manifestations of force, the primary causal explanation. Such
a view was commonplace ca. 1750; it is the central doctrine of the
physiologico-moral views of La Mettrie, Diderot and further generations of French philosophers. This conception in all its ramifications,
a decisive matter for delimiting any paradigm, fits none of the above
possibilities.

Not only does this minor instance exemplify the extreme difficulty
in characterizing precisely the full content of a given scientific community's explicit and tacit fund of orthodox attitudes and practices; it
reaffirms the diversity inherent in scientific thought and perhaps all
thought. We can attempt to fit this thought into preordained structures, or we can revel in idiosyncracies, confusions, and conflicting
doctrines. These are polar positions, and I do not see how Kuhn's
model—a "structuralism without structures," it has been called
(Piaget)—permits us to avoid tending towards one extreme or the
other. Professor Greene has reviewed our possibilities regarding the
history of natural history and has found no easy solutions. Historical
patterns, even for the moment of change which was a peculiar object
of Kuhn's *Structure of Scientific Revolutions*, may exist, but the
argument and evidence presented here might well make us cautious
to the point of timidity.

Commentary on the Paper of John C. Greene

By Leonard G. Wilson, University of Minnesota

THE DISCUSSION in both Thomas Kuhn's *The Structure of Scientific Revolutions* and Professor Greene's paper concerns the analogies and metaphors which may be appropriate to describe the process of historical change in science. In both treatments the term "paradigm" plays a central role, and therefore it may be worthwhile to consider exactly what this word means. Oddly enough, although Professor Kuhn tells us why he chose the word "paradigm," he does not define it. Of "paradigm" Kuhn says:

> By choosing it, I mean to suggest that some accepted examples of actual scientific practice—examples which include law, theory, application, and instrumentation together—provide models from which spring particular coherent traditions of scientific research.

The Oxford English Dictionary, on the other hand, defines a paradigm as "a pattern, exemplar, example" and illustrates its use in one instance by a sentence taken from Theophilus Gale's, *The Court of the Gentiles*, published in 1669. Gale wrote: "The Universe . . . was made exactly conformable to its Paradigm, or universal exemplar."

Kuhn's use of the word "paradigm" is therefore historically appropriate and seems, in fact, exactly similar to that of Gale. We might even refer to the Gale-Kuhn paradigm. In Professor Greene's paper, therefore, we are asked to consider the universal exemplars to which the universe of natural history may have been thought conformable before Darwin.

In the light of Kuhn's broader discussion of scientific change, there are several different possible descriptions of the state of natural history before Darwin. It may have been:

1. In a pre-paradigmatic stage, that is, it may have not been thought conformable to any universal exemplar.

31

2. It may have possessed a paradigm with which Darwin's theory would later prove to be incompatible.

3. Or, the concept of a paradigm or universal exemplar may simply be inappropriate to a discussion of the historical development of natural history. Natural history may simply not be conformable to a single unified exemplar corresponding in astronomy to the Ptolemaic system or the Copernican system. Certainly we do not find in natural history the connected mathematical relationships characteristic of astronomical systems. On the other hand, a paradigm may have been present in the study of natural history, without being immediately perceptible.

Professor Greene thinks that natural history did not have a paradigm until the late seventeenth century when a paradigm emerged in the work of John Ray and Joseph Pitton de Tournefort. Greene takes as his criterion of the presence of a paradigm Kuhn's definition, namely, "research firmly based upon one or more past scientific achievements that some particular scientific community acknowledged for a time as supplying the foundation for its further practice." On this basis he doubts that the ancient writings of Aristotle and Theophrastus on natural history had provided a paradigm for the seventeenth century.

The paradigm represented by Ray and Tournefort (I do not think that Professor Greene intended that it was confined to them) culminated in the systematic work of Linnaeus in the mid-eighteenth century. The systematic natural history of these men, writes Greene, "did not explain anything, but only named, classified, and described natural objects." At this point, we may question whether a mere method of study may constitute a universal exemplar or paradigm of part of Nature. Yet it may be that not the method itself, but the assumptions which lay behind the method, constituted the paradigm. At any rate, Professor Greene thinks that the paradigm accepted by Ray, Tournefort, and Linnaeus lasted until the year 1859, when Charles Darwin published his *Origin of Species*.

At the same time, he suggests the existence of a counter-paradigm, incompatible with the taxonomic paradigm of Ray, Tournefort, and Linnaeus and represented by Buffon's *Histoire Naturelle*, which attempted to depict the history of living things, including man, as part of the broader history of the physical world. Greene considers that the Buffonian paradigm, although incompatible with the taxonomic paradigm, was not a response to anomalies and contradictions within the taxonomic paradigm.

Now I should like to suggest that there are features in common between Linnaeus and Buffon which are perhaps broader and more fundamental than the ideas which separated these two men. Furthermore, I think that the single paradigm which they held in common, they shared also with Ray, with Tournefort, with Aristotle, and with Theophrastus. This was a view of the world in time, quite different from the modern view, and quite different from that held by Darwin. It was a view of the world which survived the change from the Ptolemaic to the Copernican system, which permeated the work of Kepler, Galileo, and to an intense degree, that of Newton. It was a view of the world as a created universe, which had been formed in the beginning by God, substantially in the way in which it now appeared to man.

From the Reformation onward, the European mind had been deeply influenced by the reading of the Bible. This was particularly true in England in the seventeenth century after the appearance of the King James translation. The Bible was taken to provide, not only an account of the creation of the world by God, but also an account of its history since the time of creation. The history of the world since the creation covered but a finite number of the generations of Man and comprehended no real change in the natural order since the first creation of Adam in the Garden of Eden. Thus in the seventeenth century the kinds of plants and animals living in the world then were taken to be those which God had created in the beginning, and their study, therefore, was a study of that first divine creation.

Aristotle and Theophrastus had not, so far as we know, an opportunity to read the Bible, and there have been serious doubts as to the fate of their souls as a result. However, it is clear that they viewed the universe as a fixed and stable natural order infused with divinity, and bearing the marks of its divine origin in its orderly relationships.

There was an alternative Greek view of the universe in the writings of Democritus and Epicurus, represented also in Latin in Lucretius' *De rerum natura*. This was a view of the world as a purely physical system which in the completeness of its physical activity and order was itself divine. Such a view was not congenial to the Christian church, and it was perhaps natural that in the thirteenth century Thomas Aquinas should find Aristotle's discussion of the universe more attractive.

The paradigm of a created world, which had continued unchanged during the limited period of time which had elapsed since the creation, encountered its first serious anomaly in the seventeenth century with the realization that the fossils found in rocks were the actual remains of once living animals. The force of this anomaly was felt very keenly by John Ray, who, although he realized that fossils were the remains of animals, found it difficult to reconcile this fact with his other beliefs.

"There follows," wrote Ray, "such a train of consequences as seem to shock the scripture history of the novelty of the world; at least they overthrow the opinion generally received, and not without good reason, among divines and philosophers that since the first creation there have been no species of animals or vegetables lost, no new ones produced."

It is, therefore, not surprising that the recognition of the true character of fossils should have been accompanied by the appearance of numerous theories of the earth intended to explain the origin and position of these fossils. The works of Steno, Lhwyd, Burnet, Whiston and Woodward are of this character.

Descartes' theory of the earth was somewhat different in intent. He seems simply to have wished to give a philosophical account of the events which took place at the time of creation without significantly altering the concept of the origin of the world as a single event, which had been followed by the stable and constant order familiar to the present.

Later theories of the earth—which appeared during the eighteenth century—in trying to account both for the structural features of the earth and the presence of fossils in rocks, tended to extend the succession of events which occurred at the first origin of the world over somewhat longer periods of time. But they also tended to retain the cataclysmic character of these events. They might also extend the period of time from the first creation to the present, but this extension was conceived still in very restricted terms.

Now, while Professor Greene has contrasted the paradigms of Buffon and Linnaeus, they are not really so different. While Linnaeus attempted to collect, arrange, and exhibit the species of plants and animals, each of which he considered to have been created in the beginning, he was not completely indifferent to their natural relationships. He acknowledged his system of classification to be artificial and to be an interim expedient intended to serve until the natural relationships between species, genera, and families might be worked out. As time went on, he even came to doubt whether every species had been formed in the beginning and came instead to believe that it may have been the genera which were created by God in the beginning, while the species were later variants produced by the different influences of environment.

Linnaeus speculated, too, upon the geographical distribution of plants. If, as he thought, creation must have taken place at one spot upon the earth surface, that locality must have contained all of the different kinds of habitats in which plants live. The only kind of locality, he thought, which would meet these requirements would be a mountain, because on a mountainside one may have a gradation from a tropical climate at the foot of the mountain through a succession of temperate climates on the higher slopes, to an alpine or subarctic climate near the summit. On the sides of a mountain in a tropical country, therefore, plants might be created which would be capable of later covering the whole earth. Thus, although Linnaeus seemed to accept a far more stable view of the world than did Buffon, he was also groping towards a concept of historical development.

Yet Linnaeus believed that the amount of change which had occurred since the creation was limited and relatively small, and that it had culminated in the present stable order in the natural world.

Near the end of the eighteenth century the number of new fossils discovered came to constitute such a serious anomaly in the paradigm of a fixed and unchanging natural order that some adjustment was required. This adjustment was provided principally by Georges Cuvier, who postulated that the history of the world had seen a succession of creations. However, each of these creations was static in that, after its creation, it remained unchanged until it was swept away. The succession of creations then became part of the series of events of the creation of the world as we know it—that is, of the final and permanent creation. The idea of permanence and stability in the present world order remained intact in Cuvier's theory of the earth, as it did also in other catastrophic theories of the earth.

In considering the events which lead toward the revolution in natural history produced by Darwin, Professor Greene considers these an essentially British development, and he lays stress upon different national styles of research in the early nineteenth century. While the differences between British, French and German scientists during this period are very marked, the interactions between them are also important.

For instance, one of the most British of British scientists in style and temper was Charles Lyell, but Lyell was deeply influenced by his long sojourn at Paris during the summer of 1823, when he came to know intimately Alexander Brongniart, Georges Cuvier, and Alexander Von Humboldt, representatives of both French and German natural history. Moreover, the development of Lyell's ideas occurred in the light of the numerous fossil discoveries, especially of large vertebrate fossils, made after 1800 and, in 1825, summarized for the scientific world in the third edition of Cuvier's *Recherches sur les ossemens fossiles*. Lyell was also influenced by the reading of Lamarck, for, I think, it was Lamarck who made Lyell aware of the intimate and essential nature of the relationship between a species and its environment.

When Lyell laid emphasis on extinction, it was because he knew from the emerging fossil record that many extinctions had, in fact, occurred throughout the history of the earth, and when he sought an explanation for these extinctions, he found it in the continual changes occurring in the environment. These changes occurred as a result of both geological changes and biological changes.

In the 1820's Lyell saw the succession of fossils and the repeated extinctions, in a sense, as anomalies in the world view in which he had been educated and which he still accepted. He sought to adjust his understanding of the world view to accommodate these facts, but he did not reject it until one morning in December, 1828. Then, on the south side of the harbor of Syracuse, Sicily, he found in a bed of soft blue marl, perfectly preserved fossil shells and corals belonging to species identical with those living in the surrounding Mediterranean. The striking fact about this marl bed for Lyell was that it lay beneath a formation of hard white limestone, containing only the casts and imprints of shells, which was several hundred feet in thickness. This single experience altered Lyell's whole understanding of all that he had previously learned in geology.

> In the course of my tour I had been frequently led to reflect on the precept of Descartes, "that a philosopher should once in his life doubt every thing he had been taught"; but I still retained so much faith in my early geological creed as to feel the most lively surprise, on visiting Sortino, Pentalica, Syracuse, and other parts of the Val di Noto, at beholding a limestone of enormous thickness filled with recent shells, or sometimes with the mere casts of shells, resting on marl in which shells of Mediterranean species were imbedded in a high state of preservation. All idea of attaching a high antiquity to a regularly stratified limestone, in which the casts and impressions of shells alone were discernible, vanished at once from my mind.

It was the kind of gestalt-shift mentioned by Kuhn. It remained a marked turning point in his thought, although it would take him years and the writing of many thousands of words to express all that he saw it meant for geology.

I think, therefore, that I must disagree with Professor Greene when he says that Lyell's *Principles of Geology* was not a response to a state of crisis in geological science. The crisis in geology was real and had been accumulating for years, even though Lyell may have been one of the few geologists aware of this crisis. But geology in Great Britain had been in a state of suspended judgment since the debate between the Huttonians and Wernerians early in the century, and this state of suspended judgment was closely akin to crisis.

Similarly, when Professor Greene writes of the state of natural history in Britain just before Darwin, "A revolution was impending, but it was to come from outside, not from within, the establishment." I think he underestimated the capacity of a given scientific paradigm to cherish within itself anomalous observations and contradictory theories. From 1832 until 1862 Lyell managed to believe in the fixity of species, although to others this view was in direct opposition to everything else that he held true about geology. Through the 1830's and 1840's the number of facts anomalous within a concept of fixed and stable species became alarmingly great. The rapidly increasing knowledge of the fossil record demonstrated more and more clearly that some kind of progressive development had occurred in the living world through time. The violent reaction to the publication of Robert Chamber's *Vestiges of Creation* was that of a scientific community somewhat nervous about the validity of its basic assumptions.

In conclusion, I find, in contrast to Professor Greene, that Thomas Kuhn's theory of paradigm change is to an almost surprising degree applicable to changes which occurred in natural history in the eighteenth and nineteenth centuries. In saying this, however, I must allow that its applicability depends very much on what one chooses to call, at any given time, a paradigm. And the significance of the term itself depends upon the depth with which its meaning for a given science and given period can be evoked.

The Counter-Reformation in Eighteenth-Century Science—Last Phase

By Robert E. Schofield, Case Western Reserve University

EARLY IN 1773 Antoine Lavoisier confided to his Journal his determination "to bring about a revolution in physics and chemistry."[1] In 1790, the year following the publication of his *Traité élémentaire de Chimie*, he announced that revolution as accomplished in his personal creation of a fundamentally new chemistry, based solely on induction from experiment.[2] Four years later, hardly time enough to savor the triumph he had so craved, this scientific revolutionary was executed at the hands of a different revolutionary tribunal.

These are the external events which necessarily define the limits of discussion of Lavoisier's own chemical investigations. But, because those investigations may reasonably be regarded as the resolution of a century of research, the study of them has shaped nearly every discussion of eighteenth-century chemistry as well.

With their eyes fixed firmly on Lavoisier's achievement, most historians of chemistry have proceeded backward through the eighteenth century seeking vantages from which that work might better be admired. Hence is it that histories of eighteenth-century chemistry are preoccupied with the pneumatic chemists—most of whom were Lavoisier's contemporaries—and with the phlogistic theory of combustion—though phlogiston was but a small part of one pre-Lavoisian chemical theory and combustion but one of the many questions that troubled eighteenth-century chemical theorists.[3]

But Lavoisier is not a valid guide for historical explorations. In his notorious reluctance to acknowledge his dependence on the discoveries of others, in his deliberate exclusion of history from the *Traité*, as something "which might draw aside the attention of the student," in his studied rewriting of notes and communications to conceal from publication the relation of his work to that of his contemporaries, Lavoisier is revealed as systematically obscuring the roots of his endeavor.[4] Yet only one historian, and that in a work

39

with severe chronological limitations, has significantly departed from the Lavoisier canon to investigate the milieu out of which Lavoisier's chemistry might have grown.[5] Seduced by his advertising rhetoric, inspired to hagiography by his political martyrdom, and rightly impressed by his evident influence on the development of chemistry, the standard English-language sources have accepted Lavoisier's self-image and generalized it against a post-revolutionary assessment of the entire century.

Lavoisier, we are told, was a physical chemist, who freed the eighteenth century from the malign influence of Becher and Stahl, to resume a scientific revolution postponed from the late seventeenth century and return chemistry to the correct path for science laid down by Boyle and Newton.[6]

This version of the chemical revolution confuses the nature of Lavoisier's quite genuine achievement and mistakes the whole thrust of eighteenth-century science. Truly to assess Lavoisier's accomplishments against the background of eighteenth-century chemistry, one must begin a study of that chemistry at its beginning—with the failure of Robert Boyle's corpuscular philosophy.

It had been Boyle's primary purpose to comprehend chemical phenomena within the mechanical philosophy of the seventeenth-century scientific revolution. That is, he hoped to explain the substantial forms and qualities of peripatetic chemists in terms of the primary, mechanical qualities—the sizes, shapes, and motions of fundamental particles of an otherwise undifferentiated matter.[7] All of the properties, indeed the very existences, of chemical elements or principles were to be reduced to dimension and kinematics:

> . . . whatever be the number or qualities of the chymical principles, if they be really existent in nature, it may very possibly be shown, that they may be made up of insensible corpuscles of determinate bulks and shapes, and by the various coalitions and contextures of such corpuscles, not only three or five, but many more material ingredients may be composed or made to result . . . [as] the solidity, taste, etc. of salt, may be fairly accounted for by the stiffness, and other mechanical affections of the minute particles, whereof salts consist.[8]

The so-called elements of the chemists were, therefore, nothing but secondary coalitions of primitive corpuscles which might persist through chemical reactions, reactions which, themselves, were but changes in combinations of variously sized and shaped particles.

Unfortunately, Boyle was unable to affirm that his particular mechanical explanations were the truest and best. All he could do was to demonstrate the possibility of explicating forms and qualities mechanically. The weakness of this indeterminancy was soon made obvious in varient solutions to the same problem using the same basic parameters.

Both Nicolas Lemery and Guillaume Homberg, for example, attempted to apply the mechanical philosophy to an explication of the solution of gold, but not silver, in aqua regia; silver, but not gold, in aqua fortis. Each used the basic Boyleian explanation of corrosiveness—the particles of the menstrua were of a size, shape, and agility to penetrate into the pores of metals where "like so many little wedges and levers, . . . [they] may be enabled to wrench open, or force asunder the little parts between which they have insinuated themselves. . . ."[9]

Lemery argued, in his popular *Cours de Chymie*, that in aqua regia the points of nitric acid particles are so blunted by spirit of salt as to be unable to penetrate the pores of silver, though they can penetrate the larger pores of the more malleable gold, disjoining its particles. In aqua fortis, the finer points of unblunted nitric acid can penetrate the pores of both gold and silver, disjoining the silver, but not the gold, for which they are too flexible.[10]

No doubt this is a reasonable and ingenious explanation, but not more so than that of Homberg, who argues that the points of nitric acid are too large to penetrate the fine pores of dense gold, though they can enter and disjoin silver; while in combination as aqua regia, the nitric acid particles strengthen the slender particles of spirit of salt, increasing their ability to disjoin gold by impelling them as they enter its pores.[11]

Here we see corpuscular chemistry sharing a fundamental insufficiency of kinematic corpuscularity in general. Without an independent means of discovering the parameters of size, shape, and motion, unique solutions to phenomena were impossible and *ad hoc* sizes and shapes proliferated as freely as scholastic qualities had previously done.

At the turn of the century practical chemists recognized the failure of mechanistic theory by a return, in discussing the application of their art, to pre-mechanical, materialistic concepts of chemical principles. Lemery, for example, divided his theory from his pharmaceutical practice, where he adopted the concept of elements, *sui generis*. These he defined as "substances separated and divided as far as our poor efforts are capable of doing," which were distinguished by regular properties and combined in various proportions to produce specific compounds.[12]

Clearly the development of chemistry could not stop in the inconclusive ingenuities of a Robert Boyle or the professional schizophrenia of a Lemery. Soon there were proposed two alternative paths out of impasse. One group, initially of Continental chemists and particularly those of the school of Stahl, supported a growing movement against mechanism in chemistry. Another group, especially in Britain, at first preferred a different tack. At the suggestion of Isaac Newton, kinematic corpuscularity was transformed into dynamics by the addition of inter-particulate central forces to the parameters of size, shape, and motion, and the arguments for a mechanical chemistry continued.

The transition to dynamic corpuscularity in chemistry is first manifest publicly in the Queries added in 1706 to the Latin edition of the *Opticks*. In Query 23 (which becomes Query 31 in the second English edition of 1717 and thereafter), the possibilities of the new approach are summarized.

The primitive corpuscles, in the beginning, are identical, small, solid, massy, hard, impenetrable particles which are possessed of certain powers, virtues, or forces by which they act upon one another, at a distance, to attract or repel. The variant sizes and shapes of compound corpuscles are produced by the coherence of lesser corpuscles under the action of these forces, and these compound corpuscles have their determinate forces, depending upon their sizes and shapes.

Deliquescence, composition and decomposition, dissolution, concretion, crystallization, cohesion, and congelation are among the phenomena to be explained in terms of attractive forces. Volatility and evaporation, fermentation and putrefaction, elasticity and disjunction are to be explained by repulsive forces, which succeed where attractive forces end.[13]

Finally Newton provides a hint as to the necessary independent determination of these new critical parameters: ". . . to derive two or three general Principles of motion from Phaenomena and afterwards to tell us how the Properties and Actions of all corporeal things follow from these manifest Principles, would be a very great step in Philosophy. . . ."[14] As he had derived gravitational forces from macroscopic motions, so his followers were to derive these other forces from microscopic motions.

It is true that Newton's Queries were not positively asserted as truth; it is true also that the dynamic corpuscularity of the early *Principia* and *Opticks* was severely moderated, if not denied, by the General Scholium of the second *Principia* and the aether Queries

(17–24) inserted into the second English *Opticks*. But it is equally true that the aethereal Newton had very little impact on British science until after 1740. For the first four decades of the eighteenth century most British natural philosophers, and some on the Continent as well, accepted as an epistemological directive the Newtonian imperative: from the motions find the forces, from the forces deduce the phenomena.

Around that imperative there developed a school of Newtonian dynamical chemists, commencing with a *Philosophical Transactions* paper of 1708 by John Keill. Keill's paper, "On the Laws of Attraction and Other Physical Properties" develops, in thirty theorems, some explicit explanations of the phenomena enumerated in the optical Queries, based on an attractive force between primitive particles decreasing in some greater ratio than the square. The corrosiveness of acid menstrua, for example, depends upon a force of attraction between particles of the menstruum and those of the body to be dissolved, which gives to the acid particles, rushing to fill the pores of that body, sufficient momentum to overcome the cohesive attraction of its particles.[15]

John Keill's work was elaborated by that of his friend, John Freind, whose chemical lectures, supposedly delivered at Oxford in 1704, but not published until 1709, are so permeated with the concepts of the 1706 *Opticks* as to permit little doubt that he had either seen a manuscript copy of that work before its publication or had drastically revised his lectures before their publication.

Like Keill, Freind declares his dependence upon the concept of short-range attractive forces. He reduces Keill's thirty theorems to eight propositions, "which are not bare speculation, but taken from the very nature of things." Like Keill, Freind's chemistry is purely qualitative, though he reaches a height of semi-quantitative explanation in *his* solution to the gold, silver, aqua regia, aqua fortis problem. Assigning letters a, b, d, and e to the unknown attractive forces of gold, silver, aqua fortis, and aqua regia; letters c and g to the cohesive forces of gold and silver; and letters f and r to the particle sizes of aqua fortis and aqua regia, he concludes with the general formulae: should f be greater than the pores of gold, aqua fortis can never dissolve gold however large be d; but if $(b-d)f$ exceed g, with f less than the pores of silver, aqua fortis can dissolve silver, while if $(b-e)r$ be less than g, silver will never dissolve in aqua regia, regardless of its pore size. And if $(a-e)r$ be greater than c, aqua regia will dissolve gold.[16]

Unfortunately, for all his speculative ingenuity Freind was never to assign numbers to his myriad parameters of a through f and r; nor did he ever adapt his mechanical explanations to anything in chemistry but its operations. To the fundamental question of the chemist relative to the unique and invariant nature of the substances formed by particular operations, Freind never attempted an answer. Indeed, he seems to have insisted that no regular classification of substances into chemical sub-groups was possible; all that mattered was the existence of homogeneous primitive particles and the attraction between them. These deficiencies did not, however, lessen the influence of Freind on the most distinguished theoretical and experimental British chemist of the first half of the eighteenth century, Stephen Hales, whose influential investigation of pneumatic chemistry, in the *Vegetable Staticks* of 1727 was explicitly based on Newton, Keill, and Freind.

Hales's primary theoretical contribution, derived from Newton's solitary quantitative analysis of short-range forces, was to add to Freind's attractive-force chemistry, the concept of repulsive-force pneumatics. It was Hales who saw the essential tension between attracting and repelling forces in the balance of nature:

> . . . if all the parts of matter were only endued with a strongly attractive power, whole nature would then immediately become one unactive cohering lump; wherefore it was absolutely necessary, in order to the actuating and enlivening this vast mass of attracting matter, that there should be every where intermixed with it a due proportion of strongly repelling elastick particles, which might enliven the whole mass, by the incessant action between them and the attracting particles . . . .[17]

According to Hales, these strongly repelling "elastick" particles were those of the air, which were strongly attracted to the particles of other matter. Air could, therefore, be transformed from an elastic fluid to a fixed state and become part of the cohering substance of bodies, but remained able to regain its elasticity, disjoining the bodies in resuming its earlier fluid form, thus giving those bodies their active energy. As the essential change was that from fluid elasticity to a fixed state, the essential parameter was change in volume, and this is the parameter whose measurement obsessed the mechanist, pneumatic chemists from that time onward.

In Hales one reaches the high point of a purely mechanistic chemistry—and the end point as well. Hales sounds the note of its failure in his comment of 1733 that the heat of effervescent mixtures may be owing to the intestine motions produced by the vastly great attractive and repulsive Forces of many of the Particles of Matter, near the Point of contact. "But as we cannot pry into the various Positions of these Particles, in their several Combinations, on which their different Effects depend, so it will be difficult to account from any Principle, even though a true one [for] . . . the very different Effects of effervescent Mixtures."[18]

This was a cry of disillusion shortly to be echoed by Jean Théophile Desaguliers, who staunchly resisted aethereal explanations, but confessed his inability to "shew by any experiment how big the Moleculae of Vapour must be which exclude Air from their Interstices and whether those Moleculae do vary in proportion to a Degree of Heat by increase of repellant Force in each watery Particle, or by a farther Division of the Particles into other Particles still less. . . ."[19]

Pieter van Musschenbroek also declared that though there may be several internal principles in bodies acting in different proportions at different distances, it was as yet, impossible to determine them: "It would be very difficult to decide this matter because no trials can be made upon first elements. . . . nor can we know after what manner these parts incumb upon one another, or how much solid they have, how much pore, what are their figures. Yet upon these will depend their different force of attraction."[20]

Dynamic corpuscularity appeared to be foundering on the same rock that sank kinematic corpuscularity—the inability to devise an independent measure of critical parameters. The consequences are revealed, in an otherwise undistinguished text of 1748, by Thomas Rutherforth. He proclaims his allegiance to the principles of Newtonian mechanical philosophy, but complains that attractive and repulsive forces are frequently applied to improper cases where ". . . these answers have something in them very like the occult qualities of the Peripatetics. For as they used to assign as many different . . . qualities as there are different appearances . . . so the philosophers who give these answers, introduce as many different sorts of attraction as there are bodies to be dissolved and fluids to dissolve them."[21]

The philosophy of dynamic corpuscularity was still to inspire the theoretical exuberances of John Rowning and the Abbé Boscovich, as Hales's dynamical chemistry was avowedly the guide to the later researches of Henry Cavendish and Joseph Priestley, but the tide of mechanism had already begun to recede in Britain by 1750. Hales's conviction that fixed air formed the cement for animal concretions and his belief that alkaline causticity was caused by sulphureous and elastic particles of air fixed in the lime are clearly fundamental to the investigations of Joseph Black.[22] And so his demonstration of the role of air in a wide range of compounds was fundamental in the teaching of Guillaume-François Rouelle.[23] Both Black and Rouelle, however, read Hales within a different philosophical context, that of an empirical materialism, in which phenomena were to be explained as the consequence of substances carrying, as distinguishing characteristics, the qualities of causing those phenomena in proportion to their quantities.

The resurgence of substantialized qualities as explanatory principles was led by two different sets of Continental natural philosophers, the Dutch "experimental Newtonians" and the Stahlian chemists. The Dutch successively refined the argument of their countryman, Bernardus Nieuwentyt, that heat was a substance. Boerhaave, for example, argued that heat was caused by the presence *and motion* of an imponderable fluid of fire. 'sGravesande and Musschenbroek carried that theme still further, to the theory that possession of the substance of fire was alone adequate to explain heat, and this view, transported to Britain, was ratified in quantitative detail by Joseph Black.[24]

It was in Britain also that electricity, which had resisted mechanistic explanation, was to attain a satisfying solution in the creation of an imponderable fluid. In both of these cases, the concept of conservation of substance provided the means of quantification of theory. The amount of heat, or of electrification was a measure of the amount of substance present.[25]

This materialism merged, about mid-century, with the anti-mechanism of Stahlian chemists to produce a materialistic, anti-reductionist chemistry. The Stahlians did not so much reject the mechanical philosophy as they denied its value for chemistry. Mechanistic descriptions of the figures and motions of primitive particles might be true enough, but they gave no assistance in the identification of substances or the explanation of laboratory reactions. Discussions of corpuscular shapes must remain rhetorical, when the forms of indivisible particles can not be seen by the best microscopes. Mechanical hypotheses are amusing mathematical speculations which can never account for the diversity of substances or their reactions. The chemist concerns himself only with those aggregates of particles which are sensible and combine in the laboratory. These, the Stahlians believe, form chemistry's elements (not necessarily the same as fundamental and peripatetic elements), which retain their properties in and through the various compounds they compose, elements so simple that they cannot, by any known method, be decomposed in the laboratory.

For Stahl, himself, there were some elements (or principles) such as fire and air which differ from water and the variety of earths in being instruments rather than constituents of chemical operations, but later Stahlians, such as Rouelle and Macquer, expanded the list of elements, and saw them all as selectively contributing to the composition of bodies as constituents, rather than being universally involved in all compounds either constituently or instrumentally.[26]

Compound bodies were presumed to reflect the qualities of the elements which composed them, and higher order compounds were assumed to form by the association of lesser compounds possessing the same elements. These two assumptions directed the activities of most chemists of the period. They permitted an indirect analysis by which compounds could be arranged into Classes, Orders, Genera, Species, and Varieties—with class and order based on presumed elementary composition—and from them developed the doctrine of affinities.

The notion that chemical composition and decomposition depends upon some inherent disposition of particular substances to unite in varying degrees can be traced back at least to Galen's doctrine of humoral attraction by similitude. Newton had given it mechanistic sanction by relating chemical combination to inter-particulate forces, but the first explicit table of affinities, that of the early French disciple of Stahl, Étienne Geoffroy in 1718, avoids the use of attractive forces and speaks only of "rapports."[27] There were persistent, and unsuccessful, efforts by mechanists to relate affinities or "elective attractions" to Newtonian forces—of capillarity, of cohesion, even of gravity—but the majority of chemists used affinity tables as a descriptive device for refining the taxonomy of elements and chemical substances, and as a way of avoiding dynamics in chemistry.

One sees the spread of Stahlian chemistry in changes of attitude in Britain and in France. Peter Shaw's early enthusiasm for the chemistry of Newton and Freind, expressed in commentaries on editions of Boyle and Boerhaave, shrank sharply after his translation of Stahl's *Principles of Universal Chemistry*. By 1760 both he and William Lewis, translator of the *Chemical Works* of Stahl's disciple, Casper Neuman, had insisted on the necessity of distinguishing between chemistry and the mechanical philosophy. Chemists leave to others disquisitions on primary corpuscles and concern themselves with grosser principles observed in the laboratory. Chemistry considers bodies as composed of dissimilar species of matter, defined by particular qualities which may be separated and transferred and are not subject to any known mechanical laws.[28]

Britain's most distinguished materialist chemist was Joseph Black who had learned chemistry in the course of William Cullen, an admirer of Stahl and Junker and exponent of Boerhaave's material theory of heat. Black's original chemical research had demonstrated the existence of variant kinds of air by employing the concept of conservation of substance, measured by the materialist parameter of chemical weight. Black praises Newton and describes his theory as most extensively useful in assisting in the understanding of chemical phenomena, but he roundly condemns the extravagant lengths of Freind, and, like Cullen, doubts the value of relating affinities to particle size, "as we know nothing of the figure of the small parts of bodies."[29]

"Chemical science," Black declares, "is obstructed by speculations on the causes of affinity," and particularly by "the attempts of ingenious men to explain the chemical operations by attractions and repulsions . . . . I may venture to say that no man ever got a clear and explicatory notion of chemical combinations by the help of attractions."[30] He even abandoned Cullen's use of affinity diagrams as too mechanical though, like Cullen, he used affinities in the materialist classification of chemical substances on which his explanatory efforts in chemistry were concentrated.

France's leader in Stahlian chemistry was Guillaume-François Rouelle, who, with his students, engineered the transfer of allegiance of French chemists from Boerhaave to Stahl. Rouelle taught Stahlian chemistry as a comprehensive, element-instrument theory in which phlogiston was but a single component. Like Stahl, he insisted that chemistry was an empirical, laboratory science. The existence of ultimate particles with shapes, masses, forces of attraction, etc. was admitted, but the shapes of ultimate particles were unknowable and speculation on the cause of chemical affinity was to be deplored. As it was impossible to isolate Stahlian principles, one concentrated on mixtes (those stable combinations of the elements or principles), the properties of which can be examined directly and the transfer of which can be followed by use of the balance. It was Rouelle also who introduced the discoveries of Stephen Hales into the mainstream of French chemistry, treating these as a demonstration of the constituent role of air in chemistry.[31]

With the work of Rouelle, Lavoisier's teacher, we are, at length, in a position to assess the nature of Lavoisier's achievement. Characteristic aspects of Lavoisier's work may be briefly summarized. He combined all the instrumental properties of chemical elements as qualities of the single, anomalous element, *calorique*, which becomes the vehicle of repulsion, elasticity, dimensional and state changes, the cause of fluidity and the producer of heat and light. He then abandoned imponderable *calorique* to get on with chemistry, in which the measure of weight and its changes becomes the essential feature of chemical analysis.[32]

> We may lay it down as an incontestible axiom that . . . an equal quantity of matter exists both before and after experiments, the quality and quantity of the elements remain precisely the same . . . .
> . . . the usefulness and accuracy of chemistry depends entirely upon the determination of the weights of the ingredients and products both before and after experiments.[33]

The "material proof" of the accuracy of analysis is the exact equivalence of the "whole weights of the products taken together, after the process is finished," to the weight of the original substances submitted to analysis.[34] So axiomatic is this, that Lavoisier frequently assumes weight conservation in his experiments, without measurement.

It was this principle of the conservation of substance, uniquely determined by weight, which transformed his definition of the element —as the last point chemical analysis was capable of reaching—otherwise trivially familiar to contemporary chemists, from banality to an operational concept. Lavoisier was explicitly uninterested in any discussion of elements which involved "the simple and indivisible atoms of which matter is composed," regarding such discussion as "entirely of a metaphysical nature," about something of which "it is probable we know nothing at all."[35]

He also evades discussion of the concept of affinity, though affinities are, of all branches of chemistry, the "best calculated of any part . . . for being reduced into a completely systematic body," because affinities make no part of elementary chemistry, and are, besides, a field in which a colleague is working.[36] The ingenuousness of this unique Lavoisian acknowledgment of another's preserve does not, however, conceal the use of affinity concepts in his work; concepts which, moreover, employ arguments for combination by similitude. Combustible substances, for example, "ought . . . to attract or tend to combine with each other . . . [as they] in general have a great affinity for oxygen."[37]

Lavoisier is, perhaps, best known for his inversion of the accepted explanation of combustion processes, as the addition of a constituent gas rather than the loss of a substantial principle; but he regarded his development of a new nomenclature and taxonomy of chemical substances as his major contribution to chemistry. Nor, as his repeated citations to Condillac in the preface to the *Traité* show, is this a simple empirical distinguishing of substances from one another. Language was an analytical method, distinctions were not merely metaphysical but established by nature. The names given to the elements should express the most general and characteristic quality of the substance, and, from them, the names of the compounds necessarily follow—for the nature of the compound is determined by those of its composing elements, which should be reflected in its name, in the proper proportion and degree of saturation.[38]

And hence it is that Lavoisier's most celebrated error is not a trivial mistake in nomenclature but instead reveals a basic property of his chemical system. His belief that the name and class of a substance, based upon supposed external characteristics, revealed some transcendental truths about substances led him astray with respect to the role of oxygen—i.e. the acid former—in chemical operations. It also demonstrates that, to Lavoisier, the elements are irreducible carriers of quality.

Now compare Lavoisier's chemical revolution to the chemical systems preceding. Clearly his "fundamentally new chemistry" was composed of elements of earlier systems. This is not to deny that his was a chemical revolution, nor does it denigrate his accomplishments to find that he had semi-demi-precursors. We are, however, justified in demanding a reappraisal of the essential quality of that revolution.

His is not the belated application to chemistry of a mechanical revolution previously ended at Boyle and Newton. The continuation of that revolution in chemistry had had its success in Stephen Hales and thereafter dissolved in frustration. Lavoisier was not a physical chemist, but the law-bringer to analytical chemists and his vaunted induction from experiment, though more than the usual litany of the eighteenth-century natural philosopher, is only the substitution of experimental principles commensurate with a modern age, for those previously followed, which revealed their mechanistic origin.

Far from freeing chemistry from Stahl, he ordered and rationalized Stahlian chemistry, and in doing so, changed the emphasis of future chemists' activities from a futile attempt at an overly sophisticated physical reductionism to the jig-saw puzzle problems of permutations and combinations of elements. He did not do for chemistry what Newton had done for mechanics, but what Linnaeus had done for botany. For this first phase of the scientific revolution of the eighteenth century was, in fact, the last phase of a counter-reformation in which pre-revolutionary formal qualities were materialized in substance. It was the triumphant stage of an eighteenth-century neo-Aristotelean reaction against mechanism.

1. The Journal notation is, in fact, dated 1772, but has been acknowledged to be 1773 since the work of A. N. Meldrum, *The Eighteenth-Century Revolution in Science—First Phase* (Calcutta, Macmillan [1930]), 9.

2. Lavoisier's announcement of the completed revolution, with a description of its character, can be seen in his letter to Benjamin Franklin, February, 1790, quoted by Denis I. Duveen and Herbert S. Klickstein "Benjamin Franklin (1705–1790) and Antoine Laurent Lavoisier (1743–1793), Part I. Franklin and the New Chemistry," *Annals of Science* 11 (1955), 127–28. The personal nature of the achievement is re-emphasized in Lavoisier's *Mémoires de Chimie*, probably written in 1792 and published posthumously in 1805. The theory is not, as it has been called, the theory of the French chemists, *"elle est la mienne,"* quoted, p. 106, by Henry Guerlac, "A Curious Lavoisier Episode," *Chymia*, 7 (1961), 103–108.

3. This imbalance so pervades histories of chemistry that it is nearly invidious to particularize. Examples are: Aaron J. Ihde, *The Development of Modern Chemistry* (New York, Harper & Row, 1964); Henry M. Leicester, *Historical Background of Chemistry* (New York, John Wiley & Sons, Inc., 1965); and J. R. Partington, *A Short History of Chemistry* (New York, Harper & Brothers, Harper Torchbook, 1960), third edn.

4. Lavoisier's "plagiarism" and his rewriting of papers have been argued in nearly every work about him, as, for example, Douglas McKie, *Antoine Lavoisier, The Father of Modern Chemistry* (Philadelphia, J. B. Lippincott Company, 1935), 212–13, 219, 240–44. More recently, Henry Guerlac has disclosed that Lavoisier rewrote his famous *pli câcheté* of 1772 for publication, see pp. 103–104, of his "A Curious Lavoisier Episode."

5. Henry Guerlac, *Lavoisier—The Crucial Year. The Background and Origin of His First Experiments on Combustion in 1772* (Ithaca, N.Y., Cornell University Press, 1961); see also his "Some French Antecedents of the Chemical Revolution," *Chymia*, 5 (1959), 73–112.

6. This is a view summarized from most of the leading works, but drawn primarily from: (1) the standard English-language reference book in the history of chemistry, J. R. Partington's *A History of Chemistry*, III (London, Macmillan & Co., Ltd., 1962); (2) the two most nearly satisfying full-length biographies of Lavoisier in English, Douglas McKie's *Antoine Lavoisier, op. cit.*, and his *Antoine Lavoisier, Scientist, Economist, Social Reformer* (New York, Schuman, 1952); and (3) the chapter, "The Postponed Scientific Revolution in Chemistry," in Herbert Butterfield's widely (and rightly) praised general text, *The Origins of Modern Science* (New York, The Macmillan Company, 1951).

7. For the nature of Boyle's work, see any of the relevant studies of Marie Boas Hall, a recent and particularly useful one being her edition of selections from his writings, *Robert Boyle on Natural Philosophy* (Bloomington, Indiana University Press, 1965).

8. Hall, ed., *Boyle on Natural Philosophy*, 202.

9. Hall, ed., *Boyle on Natural Philosophy*, 247–48.

10. Described by J. R. Partington, *History of Chemistry*, III, 33.

11. G. Homberg, *"Essais de Chimie," Memoires de l'Academie des Sciences* (for 1702, Amsterdam edition, 1737), 57–58; and *"Sur une dissolution d'Argent," Histoire de l'Academie Royale des Sciences* (for 1706, Paris edition, 1777), 49–50.

12. Quoted by Marie Boas [Hall] *Robert Boyle and Seventeenth-Century Chemistry* (Cambridge, at the University Press, 1958), 98.

13. Sir Isaac Newton, *Opticks: or, A Treatise of the Reflections, Refractions, Inflections, and Colours of Light* (London, W. and J. Innys, 1718), third edn. second issue, abstracted from Query 31, 370, 375–76; but verified against the Latin Query 23, in *Optice, sive de Reflexionibus, Refractionibus, Inflexionibus & Coloribus Lucis libris Tres* (London, Sam. Smith & Benj. Walford, 1706).

14. Newton, *Opticks*, 377.

15. John Keill, *"In qua Leges Attractionis aliaque Physices Principia traduntur," Philosophical Transactions* 26 (1708–1709), 97–110; transl. in Hutton, Shaw, Pearson, et al., *Philosophical Transactions Abridged* (London, 1809), 407–25.

16. John Freind, *Chymical Lectures: In which almost all The Operations of Chymistry are Reduced to their True Principles, and the Laws of Nature* (London, for Jonah Bowyer, 1712), 96–101, transl. from the *Praelectiones Chymicae* of 1709.

17. Stephen Hales, *Vegetable Staticks*, edit. by M. A. Hoskin (London, Oldbourne Book Co., Ltd., from the 1727 edition, 1961), 178.

18. Stephen Hales, *Statical Essays: containing Haemastaticks* (New York, Hafner Publishing Company, New York Academy of Medicine, *History of Medicine Series* No. 22, from the London 1733 edition, 1964), 318–19.

19. J. T. Desaguliers, *A Course of Experimental Philosophy* (London, A. Millar, 1763), third edn. (first edn., vol. 2, 1744), II, 313–14.

20. Pieter van Musschenbroek, *Elements of Natural Philosophy*, tansl. John Colson (London, J. Nouase, 1744), I, 198–204.

21. Thomas Rutherforth, *A System of Natural Philosophy* (Cambridge, for W. Thurlbourn and J. Beecroft, 1748), 8–9.

22. On animal concretions, see, e.g., *Vegetable Staticks*, 110; for the section on "particles of air" fixed in the lime, see *Vegetable Staticks*, pp. 162–63, where, however, I have "corrected" as a misprint the words "particles of fire," since Hales has repeatedly denied believing in particles of fire. As the third edition retains the word "fire," however, this is either no error or it remained unnoticed.

23. For the work of Rouelle, see Rhoda Rappaport, "Rouelle and Stahl— The Phlogistic Revolution in France," *Chymia*, 7 (1961), 73–102; and "G.-F. Rouelle: An Eighteenth-Century Chemist and Teacher," *Chymia*, 6 (1960), 68–101. In the latter paper, p. 98n, Dr. Rappaport observes that Lavoisier later used the balance which had been owned and employed by his teacher, Rouelle.

24. Boerhaave's heat theory is particularly to be found in his *Elements of Chemistry*, transl. Timothy Dallow (London, J. and J. Pemberton, I. Clarke, A. Millar and J. Gray, 1735), I, 119–54; we may note that in his chemical theory, Boerhaave attempted to maintain a Newtonian mechanism, which, however, fails in the volume on practice where, like Lemery, he returns to an empirical, descriptive materialism. Willem Jacob 'sGravesande's views

on heat are described in his *Mathematical Elements of Natural Philosophy Confirm'd by Experiment* (London, J. Senex, W. and J. Innys; and J. Osborn and T. Longman, 1726), third edn., II, 1–20; that of Musschenbroek in his *Elements of Natural Philosophy*, II, 19–57.

25. The most complete discussion of the mid-period of British natural philosophy, during the rise of imponderable fluid theories, is that of I. B. Cohen, *Franklin and Newton* (Philadelphia, American Philosophical Society, 1956).

26. The chemistry of Stahl and his school has received a bad and exceedingly scanty press. My summary of Stahlian ideas is abstracted principally from Hélène Metzger's *Newton, Stahl, Boerhaave et la Doctrine Chimique* (Paris, Librarie Félix Alcan, 1930), 93–188, but can be, at least negatively, confirmed from George Ernest Stahl, *Philosophical Principles of Universal Chemistry*, trans. Peter Shaw (London, John Osborn and Thomas Longman, 1730) where the antimechanism is revealed more by a failure to employ mechanistic explanation than by explicit denunciation. See also the French Stahlian work, Pierre Joseph Macquer, *Elements of the Theory and Practice of Chymistry* (London, A. Millar and J. Nourse, 1764), second English edn.; first French edition, Paris, 1749–51.

27. E. Geoffroy, *"Des Differents Rapports observés en chimie entre differentes substances,"* *Memoires de l'Academie Royales des Sciences* (1718), 202–12.

28. Peter Shaw, *Chemical Lectures, Publicly read at London, in the Years 1731, and 1732* . . . (London, T. and T. Longman, J. Shuckburgh, and A. Millar, 1755), second edn. corrected, 146–47; William Lewis' *Commercium Philosophico-Technicum* (1763), quoted by L. Trengrove, "Chemistry at the Royal Society of London in the Eighteenth Century—I," *Annals of Science*, 19 (1963), 191–92.

29. See Leonard Dobbin, "A Cullen Chemical Manuscript of 1753," *Annals of Science*, 1 (1936), 143, for an example of Cullen's praise of Stahl, and William P. D. Wightman, "William Cullen and the Teaching of Chemistry—II," *Annals of Science*, 12 (1956), 194, for Cullen's criticism of mechanistic chemistry and his denial of any knowledge of particle shapes.

30. Joseph Black, *Lectures on the Elements of Chemistry*, ed. by John Robison (Edinburgh, for Longman and Rees, and William Creech, 1803), I, 282–83.

31. See the papers on Rouelle by Dr. Rappaport, cited in note 23, and Henry Guerlac, "The Continental Reputation of Stephen Hales," *Archives Internationales d'Histoire des Sciences*," (1951), 393–404.

32. Chapter one, pages 1–25, of Lavoisier's *Elements of Chemistry* (New York, Dover Publications, 1965) relates primarily to *calorique*. I am initially indebted to Hélène Metzger, and particularly to her "Newton: La Théorie de l'Emission de la Lumiere et la Doctrine Chimique au XVIIIème Siècle," *Archeion*, 11 (1929), 13–25, for the interpretation of the essentially non-chemical nature of Lavoisier's element(s) of heat and light.

33. Lavoisier, *Elements*, 130–31, 297.

34. *Ibid.*, 393.

35. *Ibid.*, xxiv.

36. *Ibid.*, xxi–xxii.

37. *Ibid.*, 109.

38. *Ibid.*, lxv, 53, xxx. I am initially indebted to Charles C. Gillispie, and particularly to his unpublished paper, "Devoid of Atoms," delivered at the Conference on the History of Eighteenth Century Chemistry, Paris 1959, for pointing of the taxonomic orientation of Lavoisier's chemistry.

Commentary on the Paper of Robert E. Schofield

By Robert J. Morris, Jr., Oregon State University

PROFESSOR SCHOFIELD'S penetrating and provocative paper has two major parts: an analysis of the thrust of eighteenth-century chemistry, and an evaluation of the chemical revolution in light of that analysis. Concerning the last of these, he has placed the chemical revolution in a new and different perspective. Lavoisier's adversaries bitterly and rightly complained that the major problem with the revised nomenclature was that it required acceptance of the theory which gave it birth. Keeping firmly in mind, as I believe Schofield does, the theoretical and methodological innovations which were the bases for the revised chemical taxonomy, I believe his assessment of the nature of Lavoisier's total accomplishment is a valid one.

I disagree, however, on one point, the position of caloric in Lavoisier's chemistry. Historians have commonly characterized caloric as an anomalous substance; but for Lavoisier caloric is a substance like any other and thus must have characteristics common to matter in general. He rejected the idea that particles of caloric possess a continuous, inherent motion. He explained the subtlety of caloric as due to it and other subtle fluids comprising a fourth state of matter more rarified than vapors; caloric obeys the law of gravity even though its weight is too small to be detected; its materiality is manifest when caloric is concentrated in large quantities in a short time; and even its apparently inherent elasticity is due to some modification of the force of universal attraction.

Lavoisier could "abandon caloric to get on with chemistry" only if he ignored temperature changes and changes in fluidity or elasticity which accompany chemical reactions. Although later chemists came to consider changes of state as belonging more properly to physics than chemistry, and to accept without question that gases are merely vapors of substances with very low boiling points, Lavoisier did not and could not, as his idea that elasticity is not a primary quality but a

55

chemically produced state was a new concept and a central one for his theory. His speculations in the crucial summer of 1772 in part originated in efforts to explain the volume change necessary for air to become fixed. His solution, his "theorie Singuliere" as he called it, was that airs are really vapors and that change of state is a chemical process involving the fixation of fire matter.[1]

For Lavoisier caloric is a chemical substance, like any other, capable of entering into and separating from combination according to the laws of elective affinity.[2] His theory of combustion thus is not a simple inversion of the phlogistic explanation; combustion is a chemical reaction, as represented in a table of affinities, in which one substance replaces another in the affinity series. An increase of temperature would change the order of substances in the series, since a given arrangement is valid only at a given temperature; thus oxygen, for example, would give up caloric to join with other substances. Combustion, Lavoisier explained, is a decomposition of oxygen gas produced by some combustible substance. Using the phlogistic definition of a combustible as a substance in which the matter of fire is fixed prior to being released, oxygen gas is the chief if not sole combustible substance in nature.

Lavoisier's simple yet ingenious concept of the nature of gases explains temperature changes as well as changes of state in chemical reactions and makes these changes for the first time fully a part of chemistry, and for Lavoisier a crucial part. Here is the second sentence of his famous "Reflections on Phlogiston" read in 1785: "I have deduced all explanations from a simple principle which is that pure air . . . is composed of a particular element appropriate to it combined with the matter of fire and heat. This principle once admitted, the major difficulties of chemistry seem to dissipate and vanish, and all phenomena are explained with an amazing simplicity."[3] Wherever he presented a general account of his system, he began with a discussion of caloric; and I would suggest that he did this not so he could be done with it and get on to more important things, but because his concept of heat is the logical foundation from which the rest follows, dependent as the new chemistry is on his theory of the nature of gases and the important role gases play in the system.

As Schofield suggests at the beginning of his paper, calcination and combustion together form only a small part of the whole spectrum of chemical phenomena, and were Lavoisier's achievement limited to offering a new explanation for these processes, his place in the development of chemistry would not be the one he in fact occupies. Yet the controversy over the acceptance of his theory centered largely around the question of the existence and role of phlogiston and it is understandable that the efforts of historians have for the most part concentrated on this aspect of Lavoisier's work and the work which preceded him. We would all agree with Schofield however, that the preoccupation with phlogiston and pneumatic chemistry has distorted our view of Lavoisier and, even more, that of his predecessors.

Schofield presents us with a more balanced picture of eighteenth-century chemistry in emphasizing the positive contributions of Lavoisier's predecessors, particularly chemists of the Stahlian or Rouelleian school. He views the origins of this tradition as due to a return of chemists to a philosophy of "empirical materialism" accompanied by a reaction against the precepts of the two schools of mechanical philosophy, the kinematic and dynamic, which had attracted and seduced many chemists during the latter part of the seventeenth and early eighteenth centuries. The effect of these developments was to concentrate attention on the identifiable substances found in the chemical laboratory, their properties and their various combinations.

He has characterized the important results of these changes as first, a practical acceptance of the various characteristics of a substance as being embodied in the substance itself, persisting through chemical transformations; second, a general, and as he has shown often explicit, reluctance to speculate on the precise nature of the ultimate constituents of substances; and thus third, a rejection of attempts to derive the various characteristics from more fundamental, that is mechanical, properties of the ultimate constituents. He has perceptively described in broad outline the development of mid-eighteenth-century chemistry.

Central to his thesis is the nature of the mid-century reaction to mechanism, and it appears to me that this reaction may be described in large part as a modification of the mechanical philosophy rather than its rejection, which Schofield at times seems to suggest. I would like to point out several components of what he identifies as the thrust of eighteenth-century chemistry which seem to come from the mechanical philosophy.

The eighteenth-century concentration on empirical-experimental chemistry, which Schofield rightly emphasizes, was an integral part of the seventeenth-century mechanical philosophy. Although there is no logical connection between experiment and theoretical efforts to reduce phenomena to the primary qualities of particles, the two activities were almost invariably linked. Certainly the empirical-experimental tradition did not originate in the seventeenth century.

But the mechanical philosophy gave it an added impetus; and it was an explicit tenet of followers of the new school who, in reaction to what they considered to be the vague and useless scholastic philosophy, agreed with Lémery when he said "I deal with no opinion not founded on experience."[4]

Later chemists found unconvincing Lémery's demonstrations of the pointed nature of acid particles. However, this does not lessen the importance of experiment in his work. Rouelle, who like Stahl was clearly in the mainstream of the eighteenth-century empirical tradition, took most of his experiments from Lémery's seventeenth-century text,[5] and Boerhaave while ignoring the *theories* of Boyle and Lémery, had nothing but praise for their experimental work. The empirical-experimental tradition was an important element of the mechanical philosophy which continued into the eighteenth century and became an integral part of pre-revolution chemistry.

The same can be said of a related issue, the general reluctance of eighteenth-century chemists to indulge in the creation of theoretical systems not firmly founded on experience. Admittedly there was often considerable divergence between what was said and what was done by chemists of both centuries; Boyle's skepticism did not extend to his own concepts, and Freind failed to heed the caveats of his teacher. Eighteenth-century chemists rejected, as Newton did, the attempts to explain in detail the properties of substances and the course of chemical reaction solely by the shape, size, and motion of ultimate particles. Yet the methodological basis for the attitude responsible for the rejection of these explanations was itself a conscious part of the conceptual framework of most of those whose specific explanations were rejected.

Furthermore, Schofield clearly indicates that chemists, while rejecting mechanical reductionism, did not reject the concept that matter is indeed corpuscular, composed of particles of definite size, shape, and motion. It is true, Stahl said, that these particles have a fixed shape; it is not true, however, that we know what that shape is, and hence explanations based on imagined shapes are themselves imaginary. Nevertheless, the concept that matter is corpuscular was axiomatic in eighteenth-century chemistry.

Continental chemists directed their attacks by and large toward what Schofield calls kinematic corpuscularity, not toward the Newtonian, dynamic corpuscularity. And indeed most chemists accepted the view that the ultimate particles are acted upon by forces, the cause of which is as unknown and perhaps as unknowable as are the shapes and sizes of the particles on which they act.

Just as empiricism did not originate in but received an impetus from the mechanical philosophies, so too with affinity. Although Newton did not originate the concept, it received his blessing and that of his followers, as Schofield recognizes.

In keeping with the general sensitivity regarding speculation not firmly based on experience, and to avoid the charge they were resurrecting the occult qualities of the Peripatetics, both sentiments completely in the spirit of the mechanical revolution of the preceding century, eighteenth-century chemists took great pains to point out that "affinity" is a term which merely expresses the tendency under certain conditions for different particles to join together, and that it does not imply an inherent power residing in the particles themselves. Anti-reductionism, however, had not extinguished hopes that such reduction might be possible. By mid-century some chemists were using the attractionist language, and later, others equated affinity to universal gravitation modified at small distances by the size and shape of particles. Also, by mid-century chemists generally attributed the repulsive force to the presence of the matter of fire; and later some believed that the self-repulsive property of fire is itself caused by some modification of universal attraction.

Schofield points out that the cornerstone of the eighteenth-century empirical materialist tradition was the chemists' concentration on *chemical* elements rather than on *theoretical* elements, that is, in practice chemists treating the properties of chemical substances as inherent, permanent, and persisting through chemical transformations. This essential feature, the origins of which are in part found in the concept of substantial form, seems to me to be intimately associated with seventeenth-century atomism and even Cartesianism as it was sometimes applied. The corpuscularian belief in the existence of fixed, immutable particles, whether simple or compounded, is congruent with the view that specific physical substances persist through chemical change; and the chemists of both centuries who described processes in these terms were corpuscularians of one shade or another.

The seventeenth-century mechanical revolution contributed in a more general way to the mainstream of eighteenth-century chemistry by eventually bringing chemistry fully into the bounds of natural philosophy. A substantial segment of earlier chemical speculation had been directed toward creating new cosmologies based on chemistry, theoretical structures which accounted for everything from God's original creation of the cosmos to the cause of diseases and the transmutation of metals. This tradition declined in the latter seventeenth century and became a rarity in the eighteenth; and the anti-reductionism which Schofield describes extended to the chemical as well as the mechanical reductionists.

The acceptance by chemists of the new mechanical cosmology made superfluous what had been a major thrust of theoretical chemists in the past. The adoption of the new cosmology and the concept of the structure of matter integral to it freed the chemist to direct his energies toward the realization of more limited goals. In this sense then, the mechanical philosophy set the stage for the mid-eighteenth-century chemical tradition which Schofield outlines.

The points I have raised concerning the role of the mechanical philosophy in eighteenth-century chemistry are by no means original,[6] but they do suggest that the mechanical revolution did not exactly "dissolve in frustration" after Hales. I believe these points complement rather than contradict Professor Schofield's major thesis. Indeed he himself gives much of the evidence supporting them. He has emphasized the novelty, I the continuity. But if Lavoisier's achievement is to be viewed as the reordering of Stahlian chemistry, then surely that chemistry which he revolutionized was itself in large part a reordering of the mechanical philosophy which preceded it.

1. The manuscript containing this explanation plays a prominent role in Guerlac's argument regarding Lavoisier's activities in 1772. The manuscript is reproduced in Henry Guerlac, *Lavoisier, the Crucial Year: The Background and Origin of His First Experiments on Combustion in 1772* (Ithaca: Cornell University, 1961), 215–23.

2. Indeed, for ten years he carried on a running argument with Adair Crawford, the author of a popular and influential theory which treats heat as a mechanical agent rather than a chemical constituent.

3. Translated from Lavoisier, *Oeuvres de Lavoisier* (6 vols., Paris, Imprimerie Impériale, 1862–1893), II, 623.

4. Quoted in Hélène Metzger, *Les doctrines chimiques en France du début de XVIIe à la fin du XVIIIe siècle, Première partie* (Paris, Univ. de France, 1923), 308.

5. Rhoda Rappaport, "G.-F. Rouelle: An Eighteenth-Century Chemist and Teacher," *Chymia*, (1960), 93, note 98.

6. As Schofield suggests, with few exceptions the whole area of pre-Lavoisier chemistry is badly in need of examination. Although limited and somewhat out of date, the best general discussion of seventeenth-century chemistry still is Metzger's *Les doctrines chimiques en France*; see also Marie Boas [Hall], *Robert Boyle and Seventeenth-Century Chemistry* (Cambridge, Cambridge Univ., 1958) as well as Metzger's *Newton, Stahl, Boerhaave et la doctrine chimique* (Paris, F. Alcan, 1930) and Boas, "Structure of Matter and Chemical Theory in the Seventeenth and Eighteenth Centuries," in *Critical Problems in the History of Science*, ed. by M. Clagett (Madison, Wisc., Univ. of Wisconsin, 1959), 499–514.

Commentary on the Paper of Robert E. Schofield

By Robert Siegfried, The University of Wisconsin

WITH A GREAT effort I have resisted the temptation to offer remarks on all points in Professor Schofield's paper which of themselves deserve comment. For his discourse, it seems to me, offers so many opportunities for the critic's pleasure as to constitute an embarrassment of riches. But rejecting the very real possibility that Schofield has deliberately offered this plethora of temptations in order to distract me from my chief responsibility, I will attempt to confine my comments to his main theme as I see it.

Professor Schofield has been concerned to provide Lavoisier with an eighteenth-century intellectual ancestry that derives more from Georg Stahl, the phlogistonist, than from Robert Boyle, the mechanist. He interprets Lavoisier's major work as a rationalization and an ordering of Stahlian chemistry, as ". . . the triumphant stage of an eighteenth-century neo-Aristotelian reaction against mechanism."

Let me state at the outset that after allowing myself the freedom to interpret all the "-isms" and the many umbrella-like terms Schofield uses, I am in strong agreement with his basic position. The kind of mechanism that had looked so hopeful to Boyle, had no place in Lavoisier's system of chemistry. And the four elements and the three alchemical principles that Boyle thought he had eliminated forever, did indeed provide the immediate tradition from which Lavoisier began his chemistry, and evidences of that tradition are to be clearly found in his own revolutionary work.

But Schofield in outlining Lavoisier's ancestry paints a canvas so broad that important details lack definition. In some cases a critical step in his argument consists of no more than an unsupported assertion neatly packaged in a single sentence. Some of these points seem highly plausible, while others are exceedingly doubtful; and the majority could provide opportunity for a dozen Ph.D. dissertations that have not yet been even begun.

61

In making these general criticisms at this point, I intend primarily to reinforce Schofield's implied message that the investigation of eighteenth-century chemistry is greatly in need of a renewed and freshly conceived scholarship. Some of the inadequacies of the present treatise are no more than reflections of those large omissions in our understanding of that formative century.

But the faults of earlier scholarship are not to be blamed for all the shortcomings in this discourse; some must be charged to Schofield himself. In spite of my agreement with his central thesis that Lavoisier should be seen as reflecting a Stahlian tradition, I believe Schofield has not supported his arguments nearly as well as he ought. Indeed I believe he has in important instances confused the issue by misunderstanding Lavoisier's own work.

In the first and longest part of his treatise, Schofield has described the fate of corpuscular mechanism as it rather quickly disappeared from chemical writings of the early eighteenth century. As a formality in the structure of the study, this part is useful, but Schofield has over-emphasized its importance. As he himself admits, "The Stahlians did not so much reject the mechanical philosophy as they denied its value for chemistry." Nor am I persuaded that a significant number of historians of chemistry have ever subscribed to the view that Lavoisier's work is the lineal descendant of the mechanistic philosophy of Boyle and Newton. These strictures may reduce the drama Schofield has given us in speaking of a "counter-reformation," but he has spent more than half his paper setting up and demolishing this straw man, and allowed himself too little space for his more important purpose, that of showing how Lavoisier's work can be viewed as derived from a Stahlian tradition.

In the space of about five pages, Schofield briefly describes what is for him the central characteristic of Stahlian chemistry, namely, the "resurgence of substantialized qualities as explanatory principles." This idea has its own ancestry in the iatrochemical principles, but the most familiar eighteenth-century example is of course phlogiston. Schofield identifies two such principles in Lavoisier's writings, *caloric* and *oxygen*, and it is on his discussion of these principles that the validity of the connection between Lavoisier and the earlier tradition would seem to depend. In my opinion, Schofield's discussion of both these concepts is so seriously in error that the credibility of his central thesis is threatened.

I will speak first of caloric, which I believe is a genuine and important link between Stahlian chemistry and Lavoisier. Schofield does not elaborate on the uses Lavoisier makes of this principle, stating only that it was the embodiment of many properties, "repulsion, elasticity, dimensional and state changes, the cause of fluidity and the producer of heat and light." Schofield adds that after summarizing all these properties into one principle, Lavoisier "then abandons caloric to get on with chemistry." For two distinct reasons, I find this judgment incredible!

First, it is untrue, as a reading of the *Traité Élémentaire de Chimie* makes clear. Second, Schofield by having Lavoisier abandon caloric, throws away his own best illustration of the connection between Lavoisier and the eighteenth-century tradition of chemical principles.

Schofield admits in a footnote that the first chapter of Lavoisier's *Traité* "relates primarily to caloric," but that is the only concession he makes to the possible importance of caloric in Lavoisier's scheme of chemistry. I would argue that the chapter on caloric is first because the ideas developed there are used throughout the rest of the book; that somehow caloric played a central role in Lavoisier's thinking.

Lavoisier's chief use of caloric was as the principle of the gaseous state. For him any gas could be presumed to contain caloric, the principle of fluidity and expansibility, in precisely the same way that a phlogistonist could presume the presence of phlogiston in any combustible body. Although he used caloric to explain what we would identify today as physical changes of state, his emphasis was on chemical changes in which gases were either liberated from, or fixed into the solid state. Combustion, which seemed always to require oxygen gas, also typically liberated sensible heat. To Lavoisier the process consisted of the decomposition of the oxygen gas by the combustible's superior attraction for the ponderable basis of the gas and the consequent liberation of the previously fixed caloric as sensible heat.

All this is clearly implied in the title of Part I of his *Traité* "On the Formation and Decomposition of Aëriform Fluids,—of the Combustion of Simple Bodies, and the Formation of Acids." The general discussion of caloric and the gaseous state he gives in the first chapter is then applied in chapter five on the formation of acids, in chapter seven on the formation of metallic oxides, and in chapter nine in a full-scale discussion of experimental heats of combustion.

That he treats caloric as a chemical substance is further evident from the fact that it is included in the list of elements along with all the ponderable ones, it occurs in first position in all the affinity tables in Part II when they deal with elements commonly in the gaseous state. And caloric is conserved, as measured by the ice calorimeter, with the same rigor as are the ponderable substances as measured by weight.

Not only did Lavoisier use the concept of caloric throughout the *Traité* of 1789, the posthumously published *Mémoires de Chimie* on which he was working as late as 1793 consists almost entirely of a study of caloric. And there would be no difficulty in compiling a list of major papers before the *Traité* which are devoted to the concept of caloric before the name itself was coined. Lavoisier not only used the concept of caloric throughout his career, he used it in ways clearly consistent with the earlier use of chemical principles.

Let us turn to Schofield's illustration of oxygen as principle. Here my criticism is of the opposite sort. Where he makes too little of caloric as principle of fluidity, Schofield assumes too much for Lavoisier's use of oxygen as principle of acidity, for in spite of the name, Lavoisier's use of oxygen diluted the purity of its acid-forming property by many inconsistencies. Schofield fails to point out that although all acids contain oxygen for Lavoisier, not all oxygen-containing substances are acids. Acids are formed only when oxygen is combined with acidifiable bases, such as sulfur, phosphorus, etc. This arrangement implies that the acid properties are a result of the *combination* rather than of the oxygen alone, a direct contradiction to the older use of the chemical principle.

Lavoisier was also inconsistent in regard to oxymuriatic acid (chlorine) for though it contained more oxygen than muriatic acid and should therefore have had its acidic properties enhanced, it actually showed a weaker acidity. And oxygen in combining with the metals, produced not acids, but metallic oxides, one of whose fundamental properties was that of neutralizing, or destroying the properties of acids.

Yet according to Schofield, Lavoisier's use of chemical principles is not an historical accident, a benign though revealing carry-over of earlier ideas. Rather the concept forms the very core of his chemical thinking. Schofield identifies Lavoisier's chief contribution to chemistry as the new nomenclature and the taxonomy of chemical substances. He adds, "the names given to the elements should express the most general and characteristic quality of the substance, and from them the names of the compounds necessarily follow—for the nature of the compound is determined by those of its composing elements . . . ."

I have already illustrated by a few examples how inconsistently Lavoisier applied that relationship in the case of oxygen. Let us look briefly at the other two elements whose names Lavoisier was responsible for, to test whether Schofield's point is any better supported there. These are, of course, hydrogen and azote, neither of which Schofield mentions.

The forming of water is hardly the most general or most characteristic quality for hydrogen in spite of the name, which might just as well have been given to oxygen itself. Nor does Lavoisier hint that this property is conveyed to any of the compounds of hydrogen.

With azote the irregularity is still more obvious, for though the name means "without life," referring to its quality of asphyxiating animals immersed in it, Lavoisier at the same time recognized that it was "one of the essential constituent elements of animal bodies."

Thus it appears that while Lavoisier retained vestiges of an earlier way of explanation by chemical principles, he was not as consistent in its use as Schofield would perhaps have liked him to be. Nor has Schofield showed us very clearly what those vestiges are nor how Lavoisier used them to his own advantage. Certainly Schofield has not established his main thesis that Lavoisier's work should be viewed as the "triumphant stage of an eighteenth-century neo-Aristotelian reaction against mechanism."

In summary I would like to offer a few comments beyond the particulars of Schofield's paper. I have already spoken of the difficulty in giving an accurate and reliable account of Lavoisier and the chemical revolution because of the inadequacy of available scholarship. Historians of chemistry, rather than trying to make Lavoisier a product of the Boylean mechanical philosophy as Schofield has claimed, have more typically been concerned to make him the ancestor of the nineteenth-century chemical outlook. In their efforts to find the "father of modern chemistry" they have omitted all the eighteenth-century background that no longer looked like good science in the 1890's. Too much of this "Whig" history is evident in twentieth-century writings still.

Professor Schofield has moved toward a more balanced account by calling our attention to the need to find the intellectual antecedents of Lavoisier's thoughts and concepts in the ideas of his contemporaries and immediate predecessors. But other recent scholars with the same intent have still accepted uncritically the big picture of the chemical revolution, the primary concern with combustion and the overthrow of the phlogiston theory, and spent too much effort on detailed lines of possible influence on Lavoisier's work. Meanwhile two fundamental questions implied in the phrase "chemical revolution" remain without even a seriously attempted answer.

First, what *is* the chemical revolution? The overthrow of the phlogiston theory is hardly adequate any more, but what part of Lavoisier's work survived and formed the basis of the next period of systematic progress? This might be called the quest for Lavoisier's *permanent* contribution.

Second, what was Lavoisier's *personal* contribution; that is, what is the system of chemistry actually presented in the *Traité Élémentaire de Chimie*, and how did Lavoisier get to it? There has not been, to my knowledge, a serious analysis of the internal logic and structure of that work.

When we have better answers to these questions about the chemical revolution, we can hope for better answers to questions about its eighteenth-century antecedents.

The Energetics Controversy and the New Thermodynamics

By Erwin N. Hiebert, The University of Wisconsin

THE SIXTY-SEVENTH annual meeting of the German Society of Scientists and Physicians in Lübeck in September, 1895, provided the occasion for a heated discussion of the philosophy of energetics. Wilhelm Ostwald's lecture on "The Conquest of Scientific Materialism" was delivered at one of the general sessions[1] and became the focus of animated debates from the floor which brought many strong opinions out into the open. As the controversy gained momentum it forced both the proponents of energetics (mostly chemists) and opponents of energetics (mostly physicists) to sharpen their arguments in the light of all the available evidence—experimental and theoretical.

In discussing the historical circumstances of these scientific-philosophical quarrels, whose reverberations now are so feeble in our ears, it is well to keep in mind the fact that the nature and quality of one's opposition can influence the arguments by which opinions are supported and expressed. It was an unusual, nevertheless honest, engagement of wits between scientists who certainly were not imposters. Eventually the encounter contributed to the clarification of some basic questions about the foundations of thermodynamics within the context not only of philosophical predilection and prejudice, but also in recognition of characteristic differences in the approach to thermodynamics among chemists and physicists.

The chairman of the meetings in Lübeck was the chemist Johannes Wislicenus. Although he was Ostwald's colleague from the University of Leipzig, he was not exactly enthusiastic about Ostwald's extravagant claims for energetics. In fact, Wislicenus had asked the chemist Victor Meyer of Heidelberg to deliver a lecture in the general sessions which might offset Ostwald's strong-willed and persuasive theatrics. Meyer lectured on the "Problems of Atomistics." Georg Helm, a physical chemist from Dresden, supported Ostwald's position in a paper entitled "On the Present State of Energetics."[2]

Following these lectures, the physicists, headed by Boltzmann, sharply criticized the energeticist position. Ostwald felt that he never before had experienced such unanimous hostility. Some thirty years later, long after energeticism and anti-atomism were dead issues, Ostwald recorded his impressions of the 1895 Lübeck conference as follows: "At the discussions [in Lübeck] I found myself before a closed antagonism. My only supporter and fighting companion was G[eorg] Helm. . . . But he was removed from me because of his aversion to a realistic conception of energy . . . . It was the first time that I personally found myself confronted by such a unanimous band of downright adversaries; later on I experienced this again several times."[3] Ostwald's "later on" had reference mainly to his vehement disputes with the proponents of the atomic theory.

Georg Helm, in a letter to his wife, gives us an immediate on-the-scene reaction to Lübeck in 1895 when he writes: "The great activity lies behind me. I believe the lecture was a success. It was applauded and praised; but during the discussion there was a stiff fight. Boltzmann commenced with friendly appreciative remarks . . . and [then] began to inveigh against Ostwald's and my work. He, later Klein, Nernst, Oettingen touched matters for which I was not at all prepared. . . . Ostwald and Boltzmann came to heavy blows . . . . The meeting lasted from 9 until after 12 . . . . The hall was more than half full, so that a few hundred persons experienced the whole business. . . ."[4]

Certainly the interchange of ideas at the conference in 1895 helped to draw the attention of some prominent scientists to the urgent need for a critical re-examination of the energy and entropy concepts. By 1900, thermodynamics had gone through the first stages of a drastic purification process. The physical meaning of the so-called law of degradation of energy, the mechanical implications of temperature and entropy in relation to kinetic and probability theory, and the crucial distinction between reversibility and irreversibility all had been clarified substantially. Thermodynamics had emerged virtually liberated from its energetistic embellishments, free to contend with more important matters—such as finding a solution to the indeterminate integration constant in the generalized form of the Gibbs-Helmholtz equation, and the taking of some bold steps in the direction of axiomatization.

In 1895 an interest in the energeticist controversy was evident among persons known for their critical standards, but who nevertheless were open-minded to the discussion of various unorthodox modes of thought which then were alive. Thus, the mathematical or theoretical physicists pursuing physics in the spirit of Felix Klein notably became vocal in their challenge of the energeticists. Not given over to reckless speculation, they stood for clarity of conception and mathematical rigour. One such mathematician, Arnold Sommerfeld, was deeply impressed in Lübeck by Boltzmann's position on energetics.[5] As a young man of 27, while working on his thesis for habilitation, Sommerfeld attended the Lübeck meeting to report on his mathematical theory of diffraction. Along with other young mathematicians and the felicitous encouragement of Felix Klein, the prince of applied mathematicians, Sommerfeld found himself vigorously supporting Boltzmann's position against Ostwald.

Sommerfeld, perhaps the most unphilosophical theoretical physicist of his generation, wrote: "Helm of Dresden gave the paper on energetics. He was supported by Ostwald, and both were supported by the philosophy of Mach, who was not present. The fight between Boltzmann and Ostwald resembled, externally and internally, the fight of the bull with the lithe swordsman. But this time, in spite of all his swordsmanship, the toreador [Ostwald] was defeated by the bull. The arguments of Boltzmann broke through. At the time, we mathematicians all stood on the side of Boltzmann."[6]

Within six weeks after Lübeck 1895, the *Annalen der Physik* published Boltzmann's detailed 32-page criticism of the energetistic philosophy.[7] Boltzmann remarked in this paper that he considered many of Helm's and Ostwald's scientific contributions to be truly outstanding and that he hoped to be able to count them among his best personal friends.[8] We know, however, from the extant Boltzmann-Ostwald correspondence that Boltzmann's sharp critique of 1895 was followed by a decrease in the frequency of letters and a deterioration of the level of interchange of scientific ideas.[9]

In Boltzmann's critique, Helm and Ostwald were censured for their confused and erroneous derivations in dynamics and heat theory, their mathematical errors and inconsistencies, and above all for their narrow restriction of the entropy function to phenomena involving the dissipation of radiant energy. They also were reproached for their simplistic and uncritical acceptance of energetics as a panacea for so many unsolved problems in the physical sciences. In addition, Boltzmann obviously was perturbed deeply that the energeticists tried to claim Gibbs in their own camp. We know that Ostwald, Helm, Mach, Planck, and Boltzmann all had lauded Gibbs's approach to thermodynamics. Gibbs found it convenient to ignore them all and to mind his own scientific affairs in New Haven.

Boltzmann felt called upon to answer the energeticists because, as he expressed it, he deplored the fact that "repeatedly young people, who do not possess the necessary mathematical acuity for a successful profession in the area of theoretical physics, turn toward the domain of energetics with its promise of easy rewards."[10] He argued in his paper that the energeticists were guilty of fabricating various imaginary difficulties in theoretical physics without being mindful of the uncertainties introduced into their own excessively ambitious, yet sterile, programme. He remarked that he was unable to respond to some of Ostwald's arguments simply because he could not follow his train of thought. He accused Helm of maintaining that the entropy of a substance can increase in a cyclic process.

Boltzmann also chided Ostwald for his artificial accusation of the trend of narrow-minded, mechanistic-materialism among physicists. Professor Ostwald, he suggested, was riding an old horse; was challenging views which absolutely no longer were held among physicists. Boltzmann's criticisms, circumspectly worded to avoid the impression of hypostatizing the concepts and models of physics for all times, leaves the reader with an impression of some rather enigmatic sentiments surrounding questions concerning thermodynamics, mechanical description, and atomistics. Boltzmann wrote: "Probably no-one holds energy to be a reality any more; or believes that it has been proven without question that all natural phenomena can be explained mechanically . . . . We are much more cautious nowadays. . . . The precise description of natural phenomena, as independent as possible from all hypotheses, now generally is recognized to be most important. . . . For a long time already, even in the theory of gases, one no longer regards molecules exclusively as aggregates of material points, but rather as unknown systems determined by generalized coordinates."[11]

Boltzmann evidently was far removed from a dogmatic commitment to any particular natural philosophy, but he claimed that Ostwald categorically was shutting out mechanical and atomistic conceptions which probably had not yet run their course in physics and chemistry. He had Ostwald in mind when he wrote: "One would have to be completely enslaved by the new epistemological dogma [of energetics] in order to assert that the explanation of chemical compounds on the above hypothesis [the atomic theory] is not far removed from pure nonsense."[12]

In the long run the most persistent and influential critic of the energeticists was Max Planck. In 1895, at age 37, he was in the chair of theoretical physics at the University of Berlin. His short and terse seven-page critique, entitled "Against the New Energetics," was submitted to the *Annalen der Physik*, one month after Boltzmann's paper.[13]

He had not become involved, like Boltzmann, in the discussions on the floor in Lübeck, but submitted his criticism very soon thereafter. A recognizable reticence on his part is to be understood within the context of his uncertain intellectual rapport with men like Ostwald, Mach, and Boltzmann. Accordingly, it is significant to recognize the character of Planck's pre-Lübeck activities in thermodynamics, particularly since he at one time stood under the strong influence of Ernst Mach's ideas.[14] A brief examination of Planck's youthful preoccupation with thermodynamics is in order. Thermodynamics was his first and continuing love; in later years he referred to it as his "Lieblingsthema."[15] Indeed, Planck's preoccupation with classical thermodynamics never waned until after the turn of the century, when he had exhausted its usefulness in the study of the radiation spectrum and its energy distribution.

In 1870 in his doctoral dissertation at the University of Munich, Planck had attempted to analyze the theoretical significance of the entropy concept by way of a thoroughgoing criticism of previous formulations of the second law.[16] Most of all, he had been anxious to demonstrate rigorously that entropy, like the energy concept, could be deduced in a more direct manner than either Clausius or Thomson had done; viz., without reference to real physical or chemical processes, but rather derived ultimately from the less specific elements of our common experience. On this question Planck and Mach had much in common. Both stressed the prevalence of primitive and intuitive notions about the intimate relation of cause and effect and its connection with conservation notions. Both recognized the strength of the argument based on the impossibility of setting up a system to create an unlimited amount of work *ex nihilo*.

While Planck's views in these matters were at one with his strong attraction for generally valid basic laws underlying the knowable, real world, Mach rather stressed the primitive elements of experience and the practical demands of economy of thought as an historical determinant in the development of man's scientific views. For Planck, the energy principles occupied the place of hard-won eternal truths. Mach looked upon them as useful and conventional theoretical constructs.

Notwithstanding Mach's displeasure and irritation with anyone who called him a philosopher or spoke of his philosophical positivism, Mach at heart was an ardent philosopher puzzled by epistemological problems all his life. Not so Planck. True, like many of his scientific colleagues, he read certain philosophers. Notably in his later years he reflected on certain metaphysical issues connected with the fundamental and absolute constants of nature, causality, and realism. For Planck this realism stood in correspondence with a discoverable unity of the physical world; and virtually became a religious world view. Unlike Mach, Planck's philosophical deliberations basically were about the foundations of physics and not about science in general. In a word, Planck's philosophy was that of a naïve realist who believed in a really real world, toward the understanding of which physics was lending its guiding light. Mach's philosophy, on the other hand, cannot be categorized so easily.

In 1887, while still a Privatdozent in Munich, Planck wrote an historical and analytical essay on the principle of conservation of energy which merited the second prize of the Philosophical Faculty of Göttingen.[17] In this work Planck praised, as Mach had done earlier, the anti-metaphysical and anti-mechanistic aspects of Robert Mayer's work on the conservation of energy. While Planck, unlike Mach, supposed that mechanical theories might play a dominant role in any unitary formulation of the mode of operation of the forces of nature (Naturkräfte), he was on the side of Mach in admitting that the energy principle would survive, if necessary, without a mechanistic interpretation. That is, like Mach, Planck did not adopt an *a priori* dogmatic position about mechanical theories; but unlike Mach, he welcomed the prevalent mechanistic trends.

As an associate professor of theoretical physics in Kiel, Planck turned his attention (in four papers 1887–1891) to the application of the second law to chemistry. His motivation for doing this was to demonstrate that the second law deserved the same position of universal generality which long had been attributed to the first law.[18] His efforts in this regard were focused upon some theoretical questions and the application of the second law to the rapidly growing new areas of physio-chemical and thermochemical investigation which were just then being opened up—notably through the inspiration of men like Ostwald, van't Hoff and Arrhenius.[19]

His approach to the subject, which unquestionably pleased energeticists like Ostwald and Helm no less than it did Mach, was to proceed without paying any attention whatsoever to questions concerning the ultimate structure of matter. It was not so much that he wanted to reject outright and for all times the value of an atomic-kinetic interpretation of physical phenomena. It was rather that thermodynamics was providing successful solutions to some important scientific questions, whereas atomistics and kinetics, Planck believed, had been leading a quite sterile existence.

In September of 1891 Planck delivered a general lecture on thermodynamics before the physics and chemistry division of the German Society of Scientists and Physicians in Halle.[20] The central theme of the lecture was to review and compare the achievements of thermodynamics against those of the kinetic theory and the atomic-molecular hypothesis when applied to thermal phenomena.

Planck's preference for the thermodynamic approach was crystal clear as he strove to drive home the extraordinary successes of the second law when applied to physical chemistry. He touched upon Gibbs's use of the entropy function, Helmholtz' free energy, Duhem's thermodynamic potential, van der Waal's equation of state, entropy calculations for the dissociation of gases based upon operations with semi-permeable membranes, van't Hoff's application of the notion of osmotic pressure to solutions, and the treatment of electrical, thermal and chemical processes with the aid of the notion of ideality and the electrolytic dissociation theory. All of these matters were illustrated with examples drawn from the works of van't Hoff, Arrhenius, Ostwald, and Nernst. Singled out as a particularly convincing application was the theoretical work of Ostwald on electrolytic conductivity as a function of dilution. These formulations, Planck indicated, did not require the postulation of a mechanical conception of nature or of any hypotheses concerning the nature of heat or electricity.

By contrast, Planck suggested that while the kinetic interpretation of Joule, Krönig, and Clausius seemed to rest upon a comparatively simple, broad, and firmly established foundation, it had little to show for in the form of specific results relating heat phenomena to the nature of molecular processes. The kinetic-molecular theory, he felt, had not lived up to the expectations of its early brilliant successes and promises. He wrote: "With every attempt to build up the [kinetic] theory more elaborately, the difficulties have mounted in a serious way. Every one who studies the works of the two investigators who probably have penetrated most deeply into the analysis of molecular motions, namely Maxwell and Boltzmann, will be unable to avoid the impression that the admirable expenditure of physical ingenuity and mathematical dexterity necessary to master the problems of the day do not compare . . . with the results achieved."[21]

For Planck, the laws of thermodynamics in conjunction with the conception of ideal processes represented a great triumph of the human spirit; for he believed that it enabled the scientist to investigate areas which otherwise are closed completely to direct experimental evidence. Nevertheless, after all of Planck's good words about the splendid achievements of thermodynamics, he momentarily backed away from his analysis as if to admit that he had exaggerated the potential long-range merits of thermodynamics over those of the kinetic theory. For he suggested that, as a theory, even the second law might some day lose its status of universal validity, or at least that the boundaries of its applicability might be restricted. Mach, if not the energeticists, would have approved at this point—especially since Planck so decidedly relegated to experience the last word on all scientific matters.

In his autobiography, *Lebenslinien*, Ostwald gives the gist of a lively discussion with Planck, Boltzmann, and Hertz which took place at the meeting in Halle in 1891, i.e., four years prior to the meeting in Lübeck. Apparently, Ostwald joined Planck in defending the thermodynamic study of chemical equilibrium against Boltzmann's kinetic considerations. Boltzmann was told that his efforts in kinetics had brought forth not a single new law; that an excessive display of mathematics was needed even to derive the known relationships. According to Ostwald, Boltzmann was unable to answer this charge in any decisive way but seemed not to be defeated by the arguments. Rather, he underscored his belief in the truth of atomistics by saying: "I see no reason not to regard energy as being distributed atomistically."[22]

There is much in Planck's address that appealed to the energeticists who were present at the Naturforscherversammlung in Halle in 1891: the relative merits of the thermodynamic view over against the kinetic, molecular, mechanistic conception of physical processes; the emphasis upon the experimental basis of the first and second laws; and a kind of enigmatic silence about any deeper meaning of the entropy concept. Actually Planck then already was engaged in rethinking the status of the second law as a general thermodynamic principle. Although as careful as possible to avoid atomistic representations in his work, there is no doubt that Planck was totally unsympathetic toward the extravagant anti-mechanistic views of the energeticists.

Over a period of years the rising voices of the energeticists so repulsed Planck that eventually he decided to attack their philosophical position head-on. However, before this occurred we note that he had convinced himself on purely scientific grounds that the entropy principle was being interpreted much too narrowly by some English physicists, by Nernst, and certain unnamed close colleagues and friends of the energeticist camp in Germany.[23]

Planck seems not to have been at liberty to say anything explicitly in print against the energeticists until his intellectual breach with Ostwald and his cohorts was complete—at least, not in view of the apparent affinity between them at the meeting in Halle in 1891. In the meantime, Ostwald's *Studies on Energetics* appeared, in June, 1891, and June, 1892.[24] Now for the first time Planck was in a position to evaluate, in print, the full thrust of Ostwald's tactics for the future of the sciences—based as it was upon a radically monistic world view devoted to the systematic substitution of energy notions for phenomena involving mechanism, atomism, and the kinetic theory. Here presented in the scholarly publications of the royal scientific society of Leipzig was Ostwald's scientific Weltanschauung founded upon the new energetics—die bewusste Energetik.

The aim of Ostwald's *Studies* was to demonstrate that the reduction of physics and chemistry to mechanics had been an abortive undertaking, whereas the advances of thermodynamics were proceeding apace daily. Because of the difficulty of giving temperature a mechanical interpretation, Ostwald suggested that Gauss's absolute system of measurement, based on the factors of *mass*, space, and time be replaced by an *energy*, space, and time framework. Indeed, Ostwald rejected all *mechanical* systems, including Helmholtz' central force ideas and Hertz's mechanics. What he intended was not merely to give "energy" the same status of reality as "substance," but rather to reduce matter to one of the complexes of the energy factor.

As for the second law, it took a subordinate place in Ostwald's energy system. Entropy was a special heat capacity factor designed to treat problems dealing with what Ostwald called "a perpetuum mobile of the second order." Energy (*das eigentliche Seiende*) along with space and time, but not matter, were taken as the essential primary realities. Entropy was a derived quantity representing the dissipation of energy associated with radiation. Radiant energy was that portion of energy which was useable by virtue of not being bound to matter in the way that other forms of energy are.

In 1893 Planck gave a new interpretation of the second law and the entropy concept, although still without any outward criticisms of the energetistic views.[25] It was an attempt to state the characteristic features of the second law, derived not from any particular postulate about heat, nor from an *a priori* demonstration, but from experience with irreversible natural processes per se.

In this work Planck stood firm in the belief that there are irreversible processes for which there is no change of temperature. He emphasized that the essential meaning of the second law simply rests upon the fact that many different kinds of irreversible processes really do exist: the conduction of heat; the free expansion of gases; the freezing of super-cooled water; the condensation of super-saturated steam; explosive processes; chemical processes which proceed in finite intervals of time; in fact, every change of a system moving toward a more stable condition of equilibrium. Planck explained that the foundations of the second law would crumble if one could demonstrate a means of rendering any of these processes reversible.

Unlike the energeticists, Planck was deeply convinced that it would be impossible to represent natural phenomena in any system which failed to recognize the fundamental difference between reversible and irreversible processes. He argued that the second law provided the necessary and sufficient conditions for distinguishing reversibility and irreversibility in terms solely of the entropy values of the initial and final states of the systems involved. In light of Planck's endeavor to give an interpretation of the second law essentially in opposition to that of the energeticists it is noteworthy that Planck at this time was also moving toward a more sympathetic view on atomism.[26]

At the Naturforscherversammlung in Lübeck in 1895 the philosophy of energetics exploded out into the open. As we have seen, within several months Boltzmann, taking the lead, had published a detailed criticism of the views of Helm and Ostwald in the *Annalen der Physik*.

And now Planck could no longer keep his criticisms of energetics under constraint, even though he was not exactly on the best of terms with Boltzmann. In his paper of 1896 Planck indicated that he was not going to defend the mechanistic view of nature. Rather, he would restrict himself, at a more elementary level, to a discussion of the mathematical inadequacies of the new energetics.[27] Furthermore, he was willing to examine the energeticists' claims carefully. He felt that they had had something to offer, but that in comparison with the pretense of being able to reduce mechanics to energy considerations they were liable to fall very far short of the mark. Among scientists, Planck certainly was not the one to minimize the importance of thermodynamics. But the energeticists, he felt, were dead wrong on two counts: They overestimated the relevance of energetics for the science of mechanics; and they failed to recognize the fundamental and unique significance of the entropy function in dealing with irreversible phenomena.

The attack on energetics by Boltzmann and Planck in 1896

brought replies in the *Annalen* from both Helm and Ostwald. In answer to Boltzmann, Helm attempted to justify his energy-based derivations in dynamics by pointing out that Planck, also, had approached the problem in a similar way in his treatise of 1893.[28] Besides, Helm believed that the opponents should give the energeticists a fair chance to clear up some of their technical and conceptual difficulties. In point of fact, Helm at this time was trying to accomplish just that and was involved, besides, in writing his *History of Energetics*, in which he hoped to separate the energetic chaff from the energetic wheat for the scientific community. After the good beginnings which Planck had made, Helm was sad, he said, to see Planck join the growing forces opposing energetics—the implication being that the energeticists had once upon a time supposed that Planck took a favorable position toward their philosophy.

Ostwald's reply to Boltzmann and Planck did not come to grips at all with the technical and mathematical criticisms which had been raised. He argued that mathematical operations do not necessarily reveal physical meaning; that analytical perfection was less important than finding a philosophy of energetics which could be formulated explicitly. In any case, he objected to considering "energy" and "entropy" as mere analytical formulations, properties of matter, functions of state, or the sum of terms of the *Verwandlungsinhalt* or *Disgregation* of the constituents.[29]

Ostwald's philosophy, in fact, had not changed substantially since his *Studien* of 1891–92. He did apologize, in general, for the errors which had resulted from trying to extend his energetistic notions beyond the areas of his own competence in physical chemistry. It was, after all, a very ambitious attempt, and one should expect some unanticipated obstacles. Indeed, he felt that a good beginning had been made and that he and his colleagues were on the right track. He justified his own approach to energetics on the basis of what he had accomplished by means of it. But he promised much more for the future. He felt that Boltzmann had been too preoccupied with correcting not the errors which he (Ostwald) had committed, but the errors which Boltzmann thought that he could have committed. The one criticism which Ostwald felt that he could get by with—and here Boltzmann was quite sensitive—was to emphasize the unfruitfulness of the kinetic theory of gases in physics as compared with the fruitfulness of the energy considerations for the development of physical chemistry. At least that was Ostwald's verdict.

While Ostwald referred to Boltzmann as his "opponent," he nevertheless appreciated the fact that Boltzmann had put forth considerable effort in analyzing his scientific contributions. Besides, Boltzmann had been a close personal friend for years. By comparison, Planck now was labeled "an outspoken opponent of energetics" who, according to Ostwald, had hardly gone beyond mentioning a single example of the mis-application of mathematics to thermodynamics, viz., the example of "volume energy." Planck had dismissed "volume energy" as *"ein mathematisches Unding."* Ostwald implied that Planck might try to respond once again, but if so, to exert a greater effort in order to bring forth more substantive criticisms, and perhaps to wait for history to judge the long-range outcome of the value of energetics.

Boltzmann, of course, was ready to respond immediately to the papers of all three: Ostwald, Helm, and Planck.[30] He acknowledged both Planck's and Helm's use of the energy principle to derive the ordinary equations of motion for a system of material points. But did this, in fact, side-step atomistics; and how about thermal phenomena? As for Ostwald's preoccupation with the need to attribute a more fundamental "substantial" existence to energy than to mass, Boltzmann's reaction was one of indifference. He knew that Ostwald was not likely to change his mind in these matters.[31] He was willing to recognize Ostwald's motivations but he rather hoped that Ostwald would do the same for the atomists.

How was Boltzmann to answer Ostwald's criticism concerning the unfruitfulness of atomistics and kinetic theory? He indicated that Gibbs's approach to thermodynamics had been based on molecular representations; that Nernst's work in electrochemistry was conceived with the aid of molecular conceptions. In both cases the atomistic views had served their purpose even without entering into the final formulation. As for the kinetic theory of gases, Boltzmann wrote: "The mathematical part of gas theory primarily pursues the objectives of the development of mathematical methodology. Immediate practical application has never been the criterion of its value. This part the pure practitioner need not read, but also not criticize."

This is where the debate stood at the end of 1896. I believe it is safe to say that from then on the energeticists steadily lost ground as the views of Boltzmann and Planck broke through to physicists and chemists everywhere. Between 1895 and 1897, after the spirit and import of "Lübeck-1895" had had time to sink in, Planck reached the omega point of tolerance toward all that was connected with the extremes of energeticism and anti-mechanism. Boltzmann felt very much the same way but for different reasons. He therefore had little to do with Planck, who still was avoiding so assiduously, in his writings and lectures, any reference to the kinetic-molecular theory.

Where did Mach stand in this controversy? Planck made no explicit references to Mach at this time. His damning criticism of Mach was at the boiling point about a decade later, not in connection with energetics, but with the atomic theory and on epistemological grounds.

By 1897, in his lectures on thermodynamics, Planck reformulated, in a comprehensive and logically constructed system, the fundamentals, definitions, and major theoretical and practical consequences of the first and second laws of thermodynamics for physics and chemistry.[32] Although this work was not characterized by any major revision of earlier ideas, it did present with more clarity and internal consistency, and with appropriate extensions and some change and equilibrium to propositions in mechanics. Planck felt that thermodynamic thought. Here again he took up the position that the second law is essentially different from the first in dealing with a question not related in any way to the first law, viz., to the direction in which a process takes place in nature.[33] From the point of view of the first law, but not the second, the initial and final states of a process were seen to be completely equivalent. Planck maintained that it was quite hopeless to attempt to reduce the laws of thermodynamic change and equilibrium to propositions in mechanics. Planck felt that it was the energeticists who, by trying to use the first law to specify the direction of thermodynamical equilibrium, had befuddled the meaning of the second law.

It is noteworthy that the publication of Planck's *Vorlesungen,* with its insistence upon the crucial distinction between reversibility and irreversibility, was coincident with the beginning of his investigations on the energy spectrum of black body radiation.[34] Within a few years, this work once more completely revolutionized Planck's views on the nature of the second law. By then, Planck had been in disfavor with the energeticists for some time. Now the new investigations of the radiation spectrum also brought him much closer to Boltzmann's probability interpretation of entropy and thus finally set him in opposition to the Machian anti-atomistic, anti-mechanistic, and anti-kinetic views with which he had flirted for twenty years—much to his detriment, he lets us know.

If we had time to pursue in more detail the latter stages of the controversy, when the more exaggerated claims of the energeticists had lost their scientific cogency and rationale, we should discover that deep, fundamental, and important scientific and philosophical questions remained—or perhaps, more accurately, came to the sur-face in a conspicuous way. To be sure, we are dealing here with the historical growth of thermodynamics as a very fundamental science. For the nineteenth century it was fundamental in this sense: Along with mechanics and electrodynamics, scientists possessed here a trio of powerful disciplines which stood apart from other great nineteenth-century accomplishments in a unique way—viz., by virtue of the far-reaching deductions which can be made on the basis of a small num-ber of basic assumptions or postulates.

Classical thermodynamics, including its twentieth-century phase as related to Nernst's heat theorem (the third law of thermodynamics) was of fundamental import in its numerous and far-reaching impli-cations. But even more fundamental, and incidentally more puzzling and enigmatic, was the revolution in thermodynamic thought which the new statistical considerations encouraged through the setting of new objectives for thermodynamics. They were vastly more ambitious than the objectives of classical thermodynamics—indeed, so ambi-tious as to encompass aspects of chemical kinetics and atomistics which were only resolved in part by quantum theory, and are still under way.

On the fringe of Boltzmann's thermodynamic deliberations, and to some extent of Planck's concerns, after he saw the light, was the perennial irritation of the energeticists' blind and stubborn rejection of the relevance of the notions of probability and irreversibility to thermodynamics. The energeticists at the end of the century, and notably among them the chemists, were basking in the proud accom-plishment of having come very close to a definitive solution in their long quest for thermodynamic (i.e. energetic) criteria for the feasi-bility or the spontaneity of natural processes. Reluctant to muddy the waters of "pure" thermodynamics, acquired over some fifty years of search, they maintained a safe distance from all speculations which might distort the beauty of the abstract and postulational structure of thermodynamics. Questions concerning the kinetics and mechanism of natural processes, and the nature and internal structure of working thermodynamic substances, simply were not seen to fall within the purview of thermodynamics as the energeticists conceived it. In their view, these studies belonged either to the legitimate but unrelated consideration of the kinetics and mechanism of process, or to the illegitimate, metaphysical and fictitious play with the particularistic, substantival, and mechanistic maneuvers of the atomists.

Boltzmann, for good reasons, felt very much isolated in his thermo-dynamic position. The glib disregard of his contributions by outstanding scientists like Ostwald and Planck seemed to him to strike at the very integrity of his own most passionate concerns. This desperate isolation is revealed, in passing, in Boltzmann's *Lectures on the Theory of Gases*. The first volume of this comprehensive treatment of the subject had been submitted to the publisher in Leipzig in September of 1895, during the month of the Lübeck conference. In the preface to the second volume, writing in August, 1898, Boltzmann apparently found it necessary to explain why he should want to continue writing about a currently unfashionable gas theory. He tells his readers that the intensified attacks on the theory "rest solely upon misunderstandings, and that the role of gas theory in science is far from being played out." In his opinion "it would be injurious for science if the prevailing contemporary hostile attitude toward the theory of gases were to lead to its oblivion, for example, as it once happened to the undulatory theory [of light] because of Newton's authority." Like a voice crying in the wilderness Boltzmann speaks of being aware of the impotence with which an individual works against the tendencies of the period, but promises that he will do whatever remains in his power to contribute to the theory of gases so that not too much will have to be rediscovered when it comes back into its own again.

Boltzmann's courageous stand reminds us that the most revolutionary elements of scientific novelty can be largely obscured or buried under the avalanche of successful on-going endeavors in science. In such a case the evidence for the internal undercover workings of a period necessarily must be more or less circumstantial for the historian of science.

From all the different kinds of direct and circumstantial evidence surrounding the birthpangs of the new thermodynamics the historian of science, of course, should recognize and struggle with the arguments, goals, achievements, and failures encountered in his study of the transition from the classical conceptions of nineteenth-century thermodynamics to the modern theoretical foundations of statistical thermodynamics and its sibling—statistical mechanics. But much more than that is necessary to put in historical focus the revolutionary shift in point of view represented by the statistical approach. Well developed already by the middle of the nineteenth century, the relevance to thermodynamics of the kinetic theory of gases, with its emphasis upon the behavior of matter in terms of atoms, was not at all obvious. It was Boltzmann's great glory to have been able to know how to develop a theory well known to him so as to give entropy a statistical interpretation—and thus to combine the thermodynamical and mechanical points of view in statistical mechanics. In a way, that was what the fuss was all about already in 1895—although it is imperative to add, that the issues were then not at all so neatly drawn.

I can give no better example of a perceptive and ingenious evaluation of the difficulties surrounding the new thermodynamics than to cite Gibbs's *Elementary Principles in Statistical Mechanics*.[35] Writing in 1901, Gibbs explicitly indicated in the preface to this work how he might avoid some of the potential difficulties of the subject by narrowing down and restricting his analysis to what he considered workable. And this tells us something important about the status of thermodynamic studies at that time.

First let us mention that Gibbs located his statistical inquiry, as he says, in that "branch of mechanics which owes its origin to the desire to explain the laws of thermodynamics on mechanical principles, and of which Clausius, Maxwell, and Boltzmann are to be regarded as the principal founders." (p. viii) Nevertheless, while, as a matter of history, Gibbs attributed the origin of statistical mechanics to investigations in thermodynamics, it seemed to him "eminently worthy of an independent development, both on account of the elegance and simplicity of its principles, and because it yields new results and places old truths in a new light in departments quite outside of thermodynamics. Moreover, the separate study of this branch of mechanics seems to afford the best foundation for the study of rational thermodynamics and molecular mechanics." (p. viii)

Gibbs maintained that "while the laws of thermodynamics may be easily obtained from the principles of statistical mechanics, of which they are the incomplete expression, . . . they make a somewhat blind guide in our search for those laws." (p. ix) Thus, while Gibbs recognized all too well the extent to which the laws of classical thermodynamics might be derived from the more comprehensive, although as yet unexploited, discipline of statistical thermodynamics, he also was aware of the difficulties of establishing, in a rigorous and comprehensive way, the foundations of statistical mechanics. More pertinent, immediately at least, was Gibbs's realistic perception of how far he could venture out, with profit, in the new direction. Curiously, the limits as he saw them remind us (at a quite sophisticated level) of the same hazards and reluctances which the old energeticists dimly perceived, but which they misread and then manipulated to their own advantage in the construction of a dogmatic and sterile philosophical system. Thus we know that Gibbs was by no means ignorant of the difficulties; rather he knew how to avoid them for the time being.

Gibbs treats this aspect of his tactics as follows:

... we [will] avoid the gravest difficulties when, giving up the attempt to frame hypotheses concerning the constitution of material bodies, we pursue statistical inquiries as a branch of rational mechanics. In the present state of science, it seems hardly possible to frame a dynamic theory of molecular action which shall embrace the phenomena of thermodynamics, of radiation, and of the electrical manifestations which accompany the union of atoms. Yet any theory is obviously inadequate which does not take account of all these phenomena. Even if we confine our attention to the phenomena distinctively thermodynamic, we do not escape difficulties in as simple a matter as the number of degrees of freedom of a diatomic gas.

... Difficulties of this kind have deterred the author from attempting to explain the mysteries of nature, and have forced him to be contented with the more modest aim of deducing some of the more obvious propositions relating to the statistical branch of mechanics. Here, there can be no mistake in regard to the agreement of the hypotheses with the facts of nature, for nothing is assumed in that respect. The only error into which one can fall, is the want of agreement between the premises and the conclusions, and this, with care, one may hope, in the main, to avoid. (pp. ix–x)

Let me conclude by returning once more to Boltzmann's views— and namely, in order to draw as sharp a contrast as possible between the approaches to thermodynamics of Boltzmann and Mach. I have chosen Boltzmann because he has been recognized, correctly I believe, to be the true founder of what later became known as statistical thermodynamics. I have chosen Mach not because he embraced the energetics of his time (rather he did not do so), but because the energeticists were trying to say something in a confused way which Mach perceived clearly and was able to express without all the nonsense which accompanied the pronouncements of men like Ostwald and Helm in the area of thermodynamics.

Boltzmann, in various papers from about 1871 to 1898, struggled with the modes by which statistical mechanics, atomistic considerations, probability theory, and irreversibility could provide fundamental explanations for the laws of thermodynamics. These views were questioned both scientifically and philosophically. The Ehrenfests in their work on *The Conceptual Foundations of the Statistical Interpretation of Mechanics* of 1911 provided an excellent review of the scientific aspects of this controversy. I, therefore, will limit myself to some remarks about the philosophical criticism of Boltzmann's views—and specifically to those given in Mach's *Wärmelehre* in 1896.

According to Mach, the mechanical conception of the second law was artificial because it was interpreted, by Boltzmann, for example, as a distinction between ordered and unordered motion, and through the establishment of a parallel between the increase of entropy and the increase of ordered motion at the expense of the disordered motion. Mach says: "If one realizes that a real analogy of the entropy increase in a purely mechanical system, consisting of absolutely elastic atoms does not exist, one can hardly help thinking that a violation of the second law . . . would have to be possible if such a mechanical system were the real seat of thermal processes. . . . In my opinion the roots of this [entropy] law lie much deeper, and if success were achieved in bringing about agreement between the molecular hypothesis and the entropy law, this would be fortunate for the [molecular] hypothesis, but not for the entropy law." (p. 364)

What we have here are two extremes to choose from. There is the molecular-mechanical interpretation of Boltzmann. And there is a thermodynamics devoid of any such mechanistic conceptions, and therefore more purely postulational. The latter, I suggest, was a responsible and useful, if unnecessarily restrictive, point of view.

In our time, of course, we recognize that the kinetic-molecular-atomic hypothesis has achieved such phenomenal successes in modern physics that any attempt to interpret thermal phenomenon without kinetic and statistical ideas seems almost ludicrous. Still I believe we should recognize as historians that Mach's philosophical considerations of simplicity and postulational economy may have sounded very plausible to many scientists at the end of the century—perhaps more plausible than Boltzmann's statistical interpretation. Such questions about thermodynamics surely are no longer of any concern to the scientist. Classical thermodynamics can be absorbed within the framework of statistical thermodynamics, since molecular motions, for example, can be taken as the source of the laws of thermodynamics. But in another sense Mach was right to see that classical thermodynamics, per se, is conceivable as an autonomous discipline which needs no assistance from mechanics in a logical sense.

In any case, it is informative to examine such matters because of their intrinsic historical merit no less than because of their bearing upon philosophical issues that are being discussed today even by scientists—viz., the give and take and comparative viability, at a given stage in the development of science, of mechanical models or pictures over against postulational purification and axiomatization. I, for one, believe that both of these antipodal prototypes, as they are employed in science, stand in need of examination; and, in fact, that they can be pursued by techniques which are not fundamentally so different as to divorce the history of science from the philosophy of science. I should like to see both of them kept in focus at the same time by the historian of science.

1. See Karl Sudhoff, *Hundert Jahre Deutscher Naturforscherversammlung. Gedächtnisschrift* (Leipzig, 1922).

2. Meyer's and Ostwald's lectures appeared in the proceedings of the society: *Deutsch. Natf. Verh.*, *67*, I, 1 (1895), 95–110 and 155–68. Only a short report of Helm's lecture was published in 1895: *Ibid.*, 67, II, 1 (1895), 28–33. See also Boltzmann *"Zur Energetik"*, *Ibid.*, *70*, II, 1 (1899), 65–68; also Ostwald, *Lebenslinien* (Leipzig, 1927), II, 179–88; and Hans-Günther Körber, *Aus dem wissenschaftlichen Briefwechsel Wilhelm Ostwalds, 1*, p. 22.

3. Ostwald, *Lebenslinien* (1927), II, 180.

4. Letter dated Sept. 17, 1895. Körber, *loc. cit.*, 118–19.

5. L. Flamm, *"Nachruf: Arnold Sommerfeld,"* *Almanach, Akad. der Wiss.* Wien, *102* (1953), 351.

6. Wiener (Oesterreichische) *Chemiker Zeitung*, 47 (1944), 25. Quoted Körber p. 22. Elsewhere, Sommerfeld refers to Mach's "inclination toward energetics." Mach, he says, was cautiously negative concerning the conception of general laws and was inclined to prefer or even to limit himself to a consideration of the energy-balance of phenomena. (*"Nekrolog: Ernst Mach,"* *Jahrbuch, Akad. der Wiss.* [Munich, 1917], 66).

On the occasion of Mach's hundredth birthday Sommerfeld wrote: "[Machs] *Prinzipien der Wärmelehre* contain, besides many brilliant conceptual explanations the foundations of the later unfruitful energetics of Ostwald's style." (*Verh. Deut. physik. Ges.* [1938], 52.) In none of these references does Sommerfeld unambiguously place Mach among the energeticists. Rather he implies, correctly I believe, that the energeticists found support for their views in Mach's treatises and in the philosophy of thermodynamics expounded there.

There is no substantial basis for the oft encountered statement that Mach subscribed to the energeticists' views any more than Planck did. This accusation probably originated through guilt by association. Ostwald and Mach, it is true, were both convinced anti-atomists until about 1910. Ostwald and Mach, just as Planck, were enthusiastic about the future role of thermodynamics. But that does not make Mach an energeticist any more than Planck— who was not quite so convinced as Mach about the useless metaphysical nature of the atomic conception but considerably more dogmatic about the reality and survival value of the energy concepts.

7. Boltzmann, *"Ein Wort der Mathematik an die Energetik,"* *Ann. Phys. Chim.*, 57 (1896), 39–71.

8. *Ibid.*, 40.

9. See Körber, *loc. cit.* Letters #1 to 71.

10. Boltzmann *"Ein Wort . . . ,"* 64.

11. *Ibid.*, 64–5.

12. *Ibid.*, 66.

13. Planck, *"Gegen die neue Energetik,"* *Ann. Phys. Chim.*, 57 (1896), 72–78.

14. See Planck's remarks in *"Wissenschaftliche Selbstbiographie"* (Halle,

1945); also reproduced in vol. 3, pp. 379–84 of *Physikalische Abhandlungen und Vorträge* (Braunschweig, 1958).

15. *Ibid.*, 381.

16. Planck, *Ueber den zweiten Hauptsatz der mechanischen Wärmetheorie* (München, 1879). In *Abh. und Vorträge* 1: 1–61.

17. Planck, *Das Prinzip der Erhaltung der Energie* (Leipzig, 1887).

18. Planck, *"Ueber das Princip der Vermehrung der Entropie,"* *Ann. Phys. Chem.*, 30 (1887), 562–82; 31 (1887), 189–203; 32 (1887), 462–503; 44 (1891), 385–428; *Abh. u. Vorträge*, 1: 196–273, 382–425.

19. See e.g. Ostwald's *"An die Leser,"* in *Ztschr. physikal. Chem.*, 1 (1887), 1–4.

20. Planck, *"Allgemeines zur neueren Entwicklung der Wärmetheorie,"* *Ztschr. physikal. Chem.*, 8 (1891), 647–56; *Abh. u. Vorträge*, 1: 372–81.

21. *Ibid.*, *Abh. u. Vorträge*, 1:371.

22. Ostwald, *Lebenslinien. Eine Selbstbiographie* (Zweiter Teil, Leipzig, 1887–1905; Berlin, 1927), 187–88.

23. Planck, *"Bemerkungen über das Carnot-Clausius'sche Princip,"* *Ann. Phys.*, 46 (1892), 162–66; *Abh. u. Vorträge*, 1: 426–27.

24. Ostwald, *"Studien zur Energetik,"* Leipzig, *Math. Phys. Ber.*, 42/3 (1891), 271–88; 44 (1892), 211–37.

25. Planck, *"Der Kern des zweiten Hauptsatzes,"* *Ztschr. f. d. physikal. u. chem. Unterricht*, 6 (1893), 217–21; *Abh. u. Vorträge*, 1: 437–41.

26. See e.g., Planck, *Grundriss der allgemeinen Thermochemie* (Breslau, 1893). This work, which was published as Vol. XI in Ladenburg's *Handwörterbuch der Chemie*, contains essentially the material discussed in Planck's four papers *"Ueber Vermehrung der Entropie"* (1887–1891). Planck's *Thermochemie* was finished in the summer of 1892. In the preface, written in September, 1893, Planck thought it necessary to explain at some length why his *Thermochemie* was devoid of any atomistic representations.

27. Planck, *"Gegen die neuere Energetik,"* *Ann. Phys. Chem.*, 57 (1896), 72–78.

28. Georg Helm, *"Zur Energetik,"* *Ann. Phys.*, 57 (1896), 650, 652.

29. Ostwald, *"Zur Energetik,"* *Ann. Phys. Chem.*, 58 (1896), 155–56.

30. Boltzmann, *"Zur Energetik,"* *Ann. Phys. Chem.*, 58 (1896), 595–98.

31. Boltzmann had tried to persuade Ostwald not to publish these views in 1892. See Körber, *loc. cit.*, letters # 9 and 11.

32. Planck, *Vorlesungen über Thermodynamik*, Leipzig, 1897. (Second edition, 1905; third, 1910; tenth, by Max von Laue, 1954; eleventh, 1963).

33. *Ibid.*, 71–110.

34. Thus a series of six papers *"Ueber irreversible Strahlungsvorgänge"* was published in the *Sitzungsberichte* of the Berlin Academy between 1897 and 1900. *Abh. u. Vorträge*, 1: 493–600, 614–67.

35. J. W. Gibbs, *The Collected Works* (1928), II, vii–xii.

Commentary on the Paper of Erwin N. Hiebert

By Lawrence Badash, University of California, Santa Barbara

I SHALL EXAMINE Professor Hiebert's very interesting paper on two levels, which I hope will illuminate some of the problems involved in reconstructing a picture of science in this period. Since I view the role of a commentator as one who tries to stimulate discussion, my comments will often take the form of questions. This should not be interpreted as analogous to reviewing a book you *wished* the author had written, but simply as an attempt to draw from Professor Hiebert and his colleagues further information.

On the first level of inquiry, I wish to call attention to the subject of energetics itself. In truth, it must have been a very fascinating field, for the concept that energy, not matter, is the basic constituent of the universe is rather attractive. An interesting and integral part of the energeticists' program was the rejection of mechanical models. Conceptual model-making had been useful for many advances, but the opponents of the kinetic-molecular-mechanical approach felt that the process incorporated unsupportable and superfluous hypotheses. In place of this allegedly weak and unfruitful method for science, they proposed an economical, energy-based approach, free from hypotheses and tied only to demonstrable and measurable phenomena. When Ernst Mach declared that "purely mechanical phenomena do not exist," his arguments may have been looked upon as mere semantics. But when he deprecated the possibility of explaining all physical phenomena by mechanical ideas, by pointing out that we have no means of knowing which of the phenomena go deepest, other scientists could not dismiss his notions out of hand, if only for the intellectually honest position of being unable to prove a universal application.

It was Wilhelm Ostwald, of course, who generated the greatest reaction in his support of energetics. In his 1895 Lübeck address he argued that, following the mid-century discovery of the equivalence of various forms of energy, the three physicists who pursued the law most closely—Helmholtz, Clausius, and Kelvin—were mistaken in assuming that all of the types of energy were, in the last analysis, forms of mechanical energy.

Ostwald further maintained that energy need not have a bearer. To arguments that energy is only an abstraction, while matter is real, he replied that the contrary is true. To his mind, matter is our construct for that which is permanent in the change of phenomena. But it is energy which is the effective thing—it affects us.

And in response to the fear of the scientific materialists, as he called them, that if the conception of moving atoms is taken away from them, what means is left of forming a picture of reality, Ostwald gave the thundering biblical admonition: "Thou shalt not make unto thee any image or any likeness!"

Such attacks on the kinetic-molecular-mechanical pursuit of science could not go unanswered, and Professor Hiebert already has told us of the replies by Boltzmann and Planck. Just what was it that killed energetics? Was it the arguments of Boltzmann and Planck, or was it the increasingly clear statement of its nature by its own proponents? Professor Hiebert seems to indicate the former, but I wonder about the latter.

I wonder, too, whether the demise of energetics was due more to the sterility of its application, or to the opposition engendered by its philosophy of rejecting atomism and proclaiming the primary nature of energy. In other words, did energetics die because it didn't work or because its philosophy was repugnant?

One might even ask whether energetics died a natural death instead of being killed. Besides the successes of statistical thermodynamics, did the other century's-end advances—such as radioactivity, the electron, and ionization theory—which spoke for a mechanical view of nature simply cause energetics to wither?

Ostwald finally admitted the reality of molecules following Einstein's work on Brownian motion. Out of sheer curiosity, I wonder if Helm was ever converted.

Ostwald complained that the kinetic school of thought required an "excessive display of mathematics." This brings to mind the circumstance that, while there is much truth in the slogan that "science is international," we must nevertheless recognize that there often are strong national trends. One such trend was the model-making, kinetic view of science, which used the mathematics that Ostwald disparaged, and which was prominent in England due to the mathematical tripos examination system in the British universities. I should like to know whether there were any English followers of energetics, who would thereby have rebelled against their educational framework.

A rapid survey of the British literature shows that the energetics controversy was not extensively treated. Ostwald's Lübeck paper was translated, Helm's position was noted, and here and there one finds a brief reference to the subject. But I have found no favorable comments and only a few in opposition. One of the latter was by G. F. FitzGerald, of Dublin, who felt that Ostwald's view of science, "a sort of well-arranged catalogue of facts without any hypotheses—is worthy of a German who plods by habit and instinct. A Briton," he continued, "wants emotion—something to raise enthusiasm, something with a human interest."

Let me even expand the question of British interest, and ask how widespread was the acceptance of energetics throughout the scientific world. Ostwald himself, at Lübeck, remarked that there existed a greater unanimity about atoms in motion constituting the world than about any other theory at any time. Just who, besides Ostwald and Helm, seriously supported energetics? Was it essentially a German controversy? I am not trying to equate importance with numbers, but I am trying to assess the significance of energetics. In this connection, we are interested not in its "correctness," but in its fruitfulness and influence.

The time-table of energetics presents some curious features. Though its early origins were British, with Rankine and James Thomson, and though Mach later contributed the philosophy of thermodynamics upon which the most rigid form of energetics was built, it seems first to have appeared in the form in which we are discussing it in Helm's book of 1887. Ostwald then wrote about it in 1891 and 1892, but the subject attracted little attention until the appearance of the second edition of his text on physical chemistry in 1893. At the 1894 meeting of the German Society of Scientists and Physicians, in Vienna, a committee was appointed to report back the following year on the "actual position of energetics." This produced the Lübeck introduction by Helm and the polemic by Ostwald. Next came the counterattacks by Boltzmann and Planck, and, as Professor Hiebert has told us, the controversy reached its climax by the end of 1896, after which the energeticists steadily lost ground.

Helm published his history of energetics in 1898, but we may perhaps regard this as a death-rattle or last gasp. Boltzmann in 1899 agreed that energetics had *some* value, but naturally spoke in favor of the mechanical approach. Yet, in this twilight of energetics, it seems surprising that in three successive years, beginning in 1899, principal addresses at the British Association meetings were devoted to spirited defenses of the mechanical conception of the universe; and that as late as 1904 an article could appear in *Science* seemingly urging that not *all* ideas of molecular forces should be discarded as energetics rolls on to success.

What can we make of the status of the energetics controversy in this light? I do not think that we need depart from Professor Hiebert's evaluation of its progress. The interest shown then in the philosophical foundations of science was, I would think, not unusual in a period of extreme turbulence, namely the Second Scientific Revolution. And by the end of the century, they were not attacking energetics, but simply boosting kinetics.

From a slightly different point of view, even if we wish to argue that energetics was alive and kicking, this attention paid to it need not necessarily be regarded as a measure of the importance credited to it among the scientific community. We may compare it to Immanuel Velikovsky's *Worlds in Collision*, which a few years ago met almost universal opposition by astronomers who ridiculed it, yet was attacked and not ignored because the astronomers feared the public would accept it.

If I may use another analogy, somehow, when hearing about the efforts of the energeticists, I am reminded of the events in Germany four decades later, when *Deutsche Physik* or Nazi science was a controversial phenomenon. The energeticists, at least, were intellectually honest, but in both movements there was a certain amount of noise generated, some public curiosity, and claims of great successes. Yet, upon closer examination, the prominent theorists in each movement were few in number (Ostwald-Helm and Lenard-Stark) and without real support from prestigious colleagues. Their followers, also few in number, were largely silent in the 1890's, and incompetent party hacks in the 1930's.

In both periods there were attacks by eminent scientists upon the rationale and goals of the movements, but these, I suggest, testify more to the public interest than to the seriousness felt of their scientific significance. When all is said and done, the vast majority of German scientists under Hitler may have given the necessary lip-service to *Deutsche Physik*, but its philosophy did not change the ways in which they worked. In the nineteenth century there was no such political coercion of scientists, but the analogy I wish to push is that energetics did not alter the scientific process for almost the totality of physicists and chemists. It had virtually no influence upon their work, either in support of, or in reaction against it. Fascinating as the concept is, energetics was, as Professor Hiebert admits, a sterile discipline. It was provocative for a time, but what did it lead to?

This question of significance leads to the other level of my observations, in which I want to make some brief comments on the relationship of energetics to statistical thermodynamics. Professor Hiebert has used energetics as the vehicle for a discussion of Planck's changing kinetic-molecular-mechanical views, and to illustrate the transition from classical to statistical thermodynamics. These seem to be the major threads in his discourse.

Since the contributions of Planck and of Boltzmann are highly important, the pertinent question is the effect of energetics upon them. Planck seems to have held some views similar to Ostwald's, but was never a confirmed energeticist. Boltzmann remained solidly against such views. We may, therefore, perhaps credit energetics with sharpening Boltzmann's steady position, and with serving as a sounding board for Planck's evolving ideas. In this sense, energetics may have failed in its positive program of creating a science based on energy, and in its negative program of destroying kinetic-molecular-mechanical concepts, but it may have been useful in generating new ideas *by its opponents*. But to what extent? This is the real question.

I am frankly uncomfortable with the juxtaposition, in Professor Hiebert's paper, of energetics and statistical thermodynamics, with the implication that one led to the other. If this is true, we are not presented with evidence for the argument, nor with information about other competing influences. Was the growth of statistical ideas in science influenced by energetics?

Regarding Planck's running battle with energetics, it is probably true that the resulting clarification of his own thoughts drove him closer to Boltzmann. But this must be balanced against the effect of his long-standing disagreement with Boltzmann concerning the statistical nature of phenomena. Because Planck was convinced of the irreversibility of certain processes connected with the concepts of energy and entropy, he could not accept a statistical interpretation that would permit the possibility of reversibility. His ultimate conversion is of great historical significance, but I wonder if the role played by energetics was not a minor one. Further, when Boltzmann, in his *Lectures on the Theory of Gases*, complained of his scientific isolation, was he really speaking of the energeticists or did he have more in mind the thermodynamicists, such as Planck?

With Professor Hiebert's concluding remarks, there can be no disagreement. Though from experience we find ourselves more comfortable with mechanical model explanations, the purely postulational approach has a certain elegance and fascination. Historians of science cannot neglect the byways, as we reconstruct the highways of science, for the byways often have not only intrinsic interest, but genuine influence. I hope that my comments will stimulate discussion about the significance of energetics.

Commentary on the Paper of Erwin Hiebert

By David B. Wilson, The University of Oklahoma

MY COMMENTS on this paper are not so much about energetics in particular as about methodological questions in general. I think I can best express my basic reaction by discussing two methodological points. Let me label them for now: (1) "the historian and positivism in science" and (2) "the historian and positivism in history."

The scientific positivists in the energetics controversy are, of course, the energeticists. They are positivists in the sense that they emphasize the importance of empirical data and concepts, like energy, that can be defined directly in empirical terms. They reject the search for unobservable entities or such underlying mechanisms as are found in the atomic theory of matter and the kinetic theory of gases. Because of these characteristics, their thermodynamics is, as Hiebert puts it, more "purely postulational" than a thermodynamics dependent upon hidden, mechanical causes. The question before us, then, is: What is the historian to say about such a positivistic explanation?

Hiebert says that the positivistic explanation, as well as the one involving mechanical models, requires examination and, further, that it "can be pursued by techniques which are not fundamentally so different as to divorce the history of science from the philosophy of science." Moreover, the implication is that his discourse is a preliminary part of this historical-philosophical pursuit. After all, Hiebert tells us that "Mach was right to see that classical thermodynamics, per se, is conceivable as an autonomous discipline which needs no assistance from mechanics in a logical sense."

In contrast to this view of an interweaving of the history and the philosophy of science, I should like to emphasize the difference between them. There *are* two different activities involved here. The historian is explaining what was; the philosopher, what is. The historian's task is to understand a past period in its own context. It is to under-

93

stand what scientists thought and why they thought it, without, I believe, trying to proclaim it right or wrong. The philosopher of science, on the other hand, is trying to determine what is right— what is proper to say about scientific knowledge, what is the proper interpretation of unobservable entities, etc. There *are* two activities— the historical task of understanding past science on its own terms and the philosophical task of determining the true nature of scientific knowledge.

There are, of course, certain obvious connections between the history and philosophy of science. The historian would certainly be interested in a past scientist's own philosophy of science and its connection with his scientific activity. For his part, the philosopher is dealing with a structure of scientific knowledge constructed over a period of time. Many of the theories he analyzes are drawn from past science. The philosopher's interest in past science has much in common with the scientist's, for the scientist, like the philosopher, is properly concerned with correct views in the past. The scientist and the philosopher, that is, are interested in how past thought measures up to present scientific and philosophical convictions.

But the fact that both the philosopher and the scientist possess an interest in past science should not obscure the difference between their interest and that of the historian. When the difference is not clearly distinguished, I think that, at least for the historian, the results are likely to be unfortunate.

With this in mind, let us examine the sorts of comments that we find in this paper concerning the energetics controversy. Boltzmann, we find, is the true founder of statistical thermodynamics. During the energetics controversy, we are told, Boltzmann, "like a voice crying in the wilderness" made a "courageous stand" against the majority energeticist view. And what do we hear of the energeticists whose views did not eventually prevail? We hear of their "blind and stubborn rejection" of the relevance of certain concepts. We hear of their "dogmatic and sterile philosophical system." We hear of their "extravagant claims." We hear of "all the nonsense" contained in their statements about thermodynamics. Finally, we hear that by 1900 "thermodynamics had emerged virtually *liberated* from its energetistic embellishments *free* to contend with *more important matters*." [Italics mine.]

Now, it seems to me, the question is: If the energeticists' ideas are to be characterized with such terms as extravagant, dogmatic, sterile, stubborn, and nonsensical, how in the world did their viewpoint ever become the dominant viewpoint? Obviously it did; we are told that the anti-energeticist Boltzmann was in a "desperate" intellectual isolation. Surely, a fundamental historical question regarding the energetics controversy is: Why did energetics become such a widespread view? A satisfactory answer to such a question, I think, can be given only by seeking to explain this controversy in its own historical setting, by attempting to show why these men thought what they did, not by emphasizing the correctness or incorrectness of their thoughts. I am not saying that this paper should be devoted entirely to answering the particular historical question I have raised. I am saying, however, that after reading an historical treatment of the controversy, we should not find ourselves bewildered in regard to such a question. It should at least seem plausible that energetics could have attracted a large following.

Now, what should the historian say about positivism in science or, for that matter, about any past scientific thought? Let me summarize my views. Past science is for the historian to explain and to try to understand within its own historical context. It is not for the historian to judge whether correct or incorrect. I am quite willing to consider the formulation of an historical-philosophical method. But I am afraid that, unless the primary task of the historian is clearly distinguished from that of the philosopher or the scientist, the history of science will suffer. I am not saying that such statements as I have tried to identify in this paper are necessarily wrong. I do think, however, that they are largely irrelevant to an historical understanding and may even serve to camouflage an historical understanding. The historian, the philosopher, and the scientist may work together, but that is not to say they are doing the same thing. And confusion at this point, I think, can only lead to confused results.

My second issue of concern is positivism in history. It seems to me that the historical positivist in this case is Hiebert. Just as the energeticists relied upon their empirical, factual data and shunned attempts to get behind that data, so also does Hiebert in this paper before us. The historian's factual data is, of course, primarily the past written record. And this commentary is largely a synopsis of the written record pertaining to the energetics controversy. In the case of Planck, for example, we get a summary of the ideas presented in his 1870 dissertation, a summary of an 1887 essay, a summary of four papers written between 1887 and 1891, a summary of an 1891 lecture, of an 1893 paper, of an 1896 paper, and of his 1897 book. The same sort of thing is true for Boltzmann and Ostwald. Of the five paragraphs devoted to Gibbs, three are almost entirely verbatim quotations from Gibbs. We get an extensive treatment of what these men wrote but not a very thorough analysis of why they wrote it.

More importantly, this positivistic emphasis overshadows and distracts from the probings into underlying causes that do exist. Let me cite a few examples. First, we should expect that Planck, described as the "most influential critic of energetics," was instrumental in the eventual abandonment of energetics. Yet, we are told that in 1896 the premier energeticist, Ostwald, regarded Planck as having said little of value.

Secondly, we should expect that Mach's influence was important in the shaping of Planck's thought, for we are told that Planck "at one time stood under the strong influence of Ernst Mach's ideas." Indeed, we are told that Mach and Planck agreed on many points. But agreement alone does not prove influence. Also, we are told that Mach and Planck differed on several points. Certainly, disagreement does not prove influence.

Thirdly, we are told that chemists tended to favor energetics and physicists to oppose it. But it is not clear why this was the case. After all, energy considerations had been valuable in physics as well as in chemistry. Likewise, the atomic view of matter had been valuable in both chemistry and physics. So, why did chemists and physicists differ so dramatically in their attitudes toward energy and atoms?

Finally, exactly what was Mach's role in all this? Mach's views are mentioned often, but we receive little indication of the degree to which they were a causal factor in the events described.

Again, I am not saying that Hiebert should have answered all these specific questions that I have noted. I am saying, however, that in this paper questions like the ones I have noted are largely ignored and that when they do appear they tend to be answered incompletely or inconsistently. Furthermore, the reason for this, it seems to me, is the positivistic character of the paper. The main goal of the discourse appears to be to provide a narrative-like summary of the events of the energetics controversy. And pursuit of this goal seems to have left scant time for a non-positivistic treatment of the historical forces underlying the written record.

My question, then, is: What should the historian say about such historical positivism? As a historian, I regard a positivistic treatment of the past as an unduly restricted treatment that will yield only a minimal amount of historical understanding. Positivism in science may be defensible; after all, the alleged hidden forces in nature are somewhat remote from us. But in the affairs of men, underlying causal factors are not so remote. We are all aware of their presence in our own thought and writing, and we should expect them to have been present in the work of men like Boltzmann, Ostwald, and Planck, too. Therefore, when I read a positivistic paper like this one, I ask: Where is the insight that we demand of the historian's creativity? To be sure, there is a measure of insight supplied by having the story of the energetics controversy told in some detail. But that is mainly to give an account of the events themselves. What I am seeking, and for the most part do not find, is the insight that comes from an imaginative explanation of why the events occurred.

Let me summarize briefly the points I am trying to make. I think there are at least two misleading methodological emphases that can hinder historical research. The first is an emphasis on who was right and who was wrong. The second is an emphasis on positivistic narrative. In the paper before us, I think there is substantial evidence of both. They are hindrances that do not necessarily preclude the achievement of an historical understanding, but, as I have tried to demonstrate, they do make that achievement extremely difficult to attain.

The Car and the Road:
Highway Technology and Highway Policy

By John B. Rae, Harvey Mudd College

"THE ROAD is one of the great fundamental institutions of mankind. We forget this because we take it for granted. It seems to be so necessary and natural a part of all human life that we forget that it ever had an origin or development, or that it is as much the creation of man as the city and the laws. Not only is the Road one of the great human institutions because it is fundamental to social existence, but also because its varied effect appears in every department of the State. It is the Road which determines the sites of many cities and the growth and nourishment of all. It is the Road which controls the development of strategies and fixes the sites of battles. It is the Road that gives its frame-work to all economic development. It is the Road which is the channel of all trade and, what is more important, of all ideas. In its most humble function it is a necessary guide without which progress from place to place would be a ceaseless experiment; it is a sustenance without which organized society would be impossible; thus, and with those other characters I have mentioned, the Road moves and controls all history.

"A road system, once established, develops at its points of concentration the nerve centres of the society it serves; and we remark that the material rise and decline of a state are better measured by the condition of its communications—that is, of its roads—than by any other criterion."[1]

So states Hilaire Belloc, English writer, essayist, critic, historian. The road as he defines it might be interpreted to include all identifiable routes of trade and travel: highway, railway, waterway, and now airway. He is, however, referring specifically to the road as we ordinarily understand the term, and when we consider that until very

The topics included in Professor Rae's article have been developed in greater detail in his book, *The Road and the Car in American Life*, published by the M.I.T. Press.

recently overland travel has moved almost entirely by road, he is justified in his position.

There is ample evidence of the vital role of highway transportation in history. The Roman Empire is the best known example, but there were other great civilizations with quite sophisticated road networks: the Maurya Empire in India and the Inca Empire in South America, for instance.

Yet from ancient times until our own era, the usefulness of highway transportation was severely limited by the inadequacy of the vehicles that had to be used. Highway engineering could provide constantly better roads. Indeed, the fundamentals of modern highway engineering were established in the eighteenth and early nineteenth centuries by the French engineers of the Corps des Ponts et Chaussées and the British school best represented by Telford and McAdam. But no matter how good the roads were, as long as loads had to be carried or vehicles hauled by animal power, highway transportation was inefficient and expensive for long distances or heavy freight. In 1816, for instance, it was calculated in the United States that it cost as much to move a ton of goods thirty miles by land as it did to carry the same goods across the Atlantic.[2] Understandably, from the earliest times traffic preferred water transport if it was available, unless there were exceptional conditions.

In the nineteenth century the development of the railroad caused an almost complete eclipse of highway transportation, and with it a deterioration of road systems, since it was no longer important to maintain roads except for strictly local traffic. The decline seems to have been worse in the United States than elsewhere; there is overwhelming evidence that American roads of the late nineteenth century were unbelievably bad.

Yet techniques of highway construction were not being forgotten; on the contrary they were advancing. The steam-roller, developed in Great Britain, provided the first really effective method of compacting road surfaces, and new materials were also being introduced experimentally. Brick and asphalt surfacing began to be used on city streets in the 1870's, and the first concrete pavement in the United States was laid around the courthouse in Bellefontaine, Ohio, in 1893.[3]

For highway transportation again to play a significant role a new vehicle technology was required that would transcend the limitations of the horse-drawn carriage and wagon. This technology was provided by two European inventions, the bicycle and the motor vehicle. The bicycle not only contributed several features to the development of the automobile, such as the pneumatic tire, but more important, it got people out on the roads in numbers and demonstrated both the possibilities of flexible personal mobility and the inadequacy of existing highways. The bicyclists, both in Europe and America—they were organized in this country as the League of American Wheelmen —were the first major segment of public opinion to agitate for improved roads toward the close of the nineteenth century.[4]

The bicycle, however, offered no solution to the problem of moving freight economically by road, and legislative bodies were not disposed to spend money on highway improvement merely for the enjoyment of bicyclists. There was pressure to "get the country out of the mud," but what was meant was local "farm-to-market" roads to assist the farmer to get his products to town, or to the nearest rail connection. The immediate result was the establishment of the Office of Road Inquiry in the Department of Agriculture in 1893, which was also the year that the Duryea car, the first American gasoline automobile, appeared.

Thus the first gestures in the direction of a comprehensive highway program coincided with the arrival of the motor vehicle, but for some years the two had little relationship to each other, except that the motorists constituted a growing body of opinion pressing for better roads.

There was ample scope for improvement. In 1904 the United States had 2,151,507 miles of highway, exclusive of city streets, of which 153,662 miles, about 7 per cent, were classified as "improved."[5] "Improved" meant any surface other than plain dirt; the roads thus classified included 38,662 miles of water-bound macadam, 108,233 of gravel surface, and the rest varied—sand, shell, even some survivals of plank roads.

At the beginning of this century, therefore, most of the existing American highway mileage needed replacing or rebuilding, so that there was an open opportunity to create a new road system adapted to the needs of an automotive age. To have done so, however, would have required not only superhuman foresight but also a body of knowledge about highway and traffic engineering that simply did not exist at that time. There was nothing in all previous experience of highway travel to suggest how roads should be designed for large numbers of fast-moving self-propelled vehicles, and if it seemed that reconstructing existing roads would take care of the problem, this was a reasonable conclusion under the conditions of the opening years of the twentieth century.

Between 1910 and 1915 automobile registrations in the United

States leaped from just under half a million to almost two and a half million,[6] and this phenomenon, together with the continuing agitation for road improvement to aid the farmers, created irresistible pressure for a national highway policy. But it was still uncertain whether such a policy should aim at a network of arterial highways or simply provide assistance for "farm-to-market" roads. One school of thought was exemplified in Carl G. Fisher's Lincoln Highway Association, created in 1913 to promote a hard-surfaced, adequately-marked coast-to-coast highway. The other was reflected in a statement made in Congress during the debate on the Federal Aid Road Act in 1916: "The railway station is the terminus for roads; neither freight nor passengers will ever be carried long distances over roads as cheaply as they could be over railways, and it is an idle dream to imagine that auto trucks and automobiles will take the place of railways in the long-distance movement of freight or passengers."[7]

The act as passed left the issue unresolved. It appropriated $75 million to be spent under the direction of the Secretary of Agriculture over a five-year period for the improvement of rural post roads.[8] Federal subsidies, in other words, were not specifically restricted to farm-to-market roads, but neither was there any suggestion of developing a network of trunk highways. The money was to be spent through state highway departments—an important provision because it compelled states without highway departments to organize them. The federal funds were granted on a matching basis, up to 50 per cent of the cost of road improvement on projects approved by the Bureau of Public Roads, but not at that time to exceed $10,000 per mile.

The Federal Aid Road Act of 1916 was recognition of the fact that good roads were essential to the national welfare and that highway improvement was therefore a national as well as a local responsibility. Before it had time to become effective the United States became involved in the First World War, so that the projected program could not go into operation under the conditions for which it had been planned. The war, however, made it abundantly evident that the country urgently needed a coherent network of trunk highways and not just a piecemeal improvement of local roads. The railroad system, especially in the Northeast, was almost brought to a halt by monumental traffic congestion. Freight cars piled up at the seaports, waiting to be unloaded, and others had to be stacked on sidings and in yards waiting their turn to move up, until the government had to take over the railroads in order to avert the threatened paralysis.

When this situation materialized, the country had to face the fact that there was no alternative system of transportation capable of providing relief for the railroads. The dependence on rail transport was such that trucks destined for the war fronts in Europe were normally shipped by rail from factory to seaport, and when the government, late in 1917, undertook to take some of the pressure off the railroads by moving some trucks from Detroit to Baltimore under their own power, a contemporary engineering journal called the experiment a "daring adventure."[9] It seemed impossible for these heavy vehicles to cross the Alleghenies on unimproved roads in winter, but they did it, in a demonstration of what highway transportation was capable of even under adverse conditions.

The war scarcely interrupted the rapid increase in numbers of motor vehicles. Registration rose from the two million of 1915 to ten million in 1921. A great expansion of highway construction was obviously called for, and if the country was to be given a coherent road network, the federal government would have to take the lead. This action came in the Federal Highway Act of 1921, which provided that federal aid should be concentrated upon "such projects as will expedite the completion of an adequate and connected system of highways, interstate in character."[10] Each state was required to designate 7 per cent of its road mileage as "primary," and this mileage alone was eligible for federal aid, matching state funds on a fifty-fifty basis. The first appropriation under this law was $75 million for the fiscal year 1922, a significant comparison with the provision made in 1916 to spend the same amount over five years.

The 7 per cent figure came to approximately 200,000 miles of road, and by 1923 the Bureau of Public Roads had planned a tentative network of arterial highways, serving every city of 50,000 or more.[11] In another two years a satisfactory mechanism for securing cooperation among the states was created, and the country's first genuine national highway system, which would eventually total some 250,000 miles,[12] was in progress. These roads also initiated the uniform method of route marking that became universal in American highway practice—even numbers for east-west routes, odd for north-south.

The Federal Highway Act of 1921 marked the beginning of intensive road-building activity at all levels of government. The significant result was qualitative rather than quantitative. Total road mileage showed a surprisingly low rate of expansion: 3,160,000 miles in 1921 to 3,690,000 in 1965, an increase of one-sixth in forty-four years.[13] Mileage of improved road was another story. From 447,000 in 1921 it reached 1,721,000 in 1945 (surpassing the mileage of un-improved road for the first time), and 2,776,000 in 1965. These figures have to be treated with some caution, because "improved" or "surfaced" in the highway statistics includes soil, slag, gravel, or stone surfacing and these accounted for a million of the surfaced miles in 1945 and half the total for 1965. They were, however, com-pacted surfaces with adequate drainage, so as to be passable in most weathers.

This qualitative advance, however, was far less effective than it might have been in providing a road system capable of permitting a rapidly increasing volume of motor vehicle traffic to flow smoothly, essentially because we were slow to appreciate that we were dealing with a completely novel set of conditions in highway transportation. Compared with the vehicles of the past, automobiles were not only much faster, they were heavier and imposed much more wear on road surfaces.

Had this been the only problem, it would have been fairly easy to solve. Highway engineers were already familiar with materials for hard surfaces, and further study and experience produced sufficient knowledge of what materials to use in given conditions. By 1924 there were 31,188 miles of concrete road, increasing at about 6,000 miles a year, and about 55,000 miles of asphalt or bituminous macadam, increasing at about 40,000 miles a year.[14]

It was more difficult to realize that there is a fundamental difference between a road system built so that motor vehicles can use it, and one specifically designed for automobile traffic. Until the motor vehicle appeared in quantity, it was no great problem for the same road to accommodate both through and local traffic, even if it was a main highway. Movement was low-speed, and, except in the center of large cities, low-density. The automobile reversed this situation and made it desirable, indeed essential, to differentiate between types of road use and to separate streams of traffic. In other words, trunk highways needed to be distinguished from local roads, and roads carrying local traffic from those intended simply to give access to property.

The primary road system provided for in 1921 was a partial recognition of this principle, but only partial. As the primary network expanded it consisted for the most part of conventional two-lane construction, hard-surfaced so that it was a marked improvement over what it replaced, but open to through and local traffic indis-criminately and with no limitation of access. Some multi-lane roads began to appear in heavily-traveled areas after 1924. These were not, as a rule, divided highways. Some were three-lane roads, a design that may be conceded to be probably the worst of all techniques for increasing capacity; the middle lane was universally designated "slaughter alley."

Second-guessing is an occupational hazard from which few historians are immune. In most respects it is probably unrealistic to suggest that the highway programs of the 1920's and 1930's could have been more effectively adapted to the requirements of motor vehicle transportation. Highway planners and engineers had to learn by experience, and they had to accumulate the experience to begin with; there was, to repeat, very little in the previous history of roads to provide guidance for an automotive age. The construction techniques and heavy road-building equipment in use today were not available then to facilitate the completion of major highway projects.

On the other hand, some of the principal components of roads suitable for automobile traffic were known. The divided roadway existed in some European and American cities, and in 1906 a French engineer, Eugene Henard, suggested grade separation of roadways, with the cloverleaf design for interchanges.[15] This concept was certainly familiar in the United States, because in 1916 Arthur Hale, of Maryland, secured an American patent on a design for a grade separation, and interchanges similar to Henard's. The first such interchange in the United States was the Woodbridge Cloverleaf in New Jersey, built in 1928 at the intersection of State Routes 4 and 25.[16] Henard also proposed the traffic circle, simultaneously with a British engineer, Holroyd Smith, and the pioneer American traffic expert, W. P. Eno, who suggested it for Columbus Circle in New York City.

The concept of the limited-access divided highway emerged during the 1920's, although it is difficult to say just when, or by whom, it was first definitely formulated. Urban parkways existed in both European and American cities as examples of special-purpose roadways. In 1924 Belloc proposed arterial highways for Britain, to be of great width, gradual curves, and no crossings at grade.[17] Later in the decade, construction of genuine express highways, the *autostrade*, was begun in Italy. To meet the heavy cost, which was the major obstacle to the building of such roads, the Italian government revived the turnpike system of former days and made the *autostrade* toll roads.

It would therefore have been technically possible for arterial roads in the United States to have been constructed as limited-access express highways by the late 1920's, and certainly in the 1930's. And even though such roads would unquestionably have been below the standards of present-day freeways, they would have aided materially in handling the accelerating volume of automobile traffic. There is no evidence that such a policy was seriously considered. It undoubtedly seemed prohibitively ambitious and expensive for a country with tremendous land distances to cover, especially when the Depression arrived and hard-pressed state governments found themselves diverting highway revenues to other purposes. Systematic study would have shown that the actual mileage required was a miniscule proportion of the country's total of over three million miles—the present Interstate system covers 41,000 miles—but highway research and planning did not become general practice with highway departments until after 1935.[18]

There was also a strong political reason for pursuing a different policy. As the federal-aid primary road system expanded, to quote an eminent American historian, "Its trunk routes passed near to the homes of most Americans, yet touched only a small fraction of the three million miles of roads. The backwardness of these was emphasized as the farmer's car pushed through the mud and then speeded up on the hard surface of the arterials."[19] Public authorities were accordingly under strong pressure to devote funds to secondary rural roads.

Consequently, during the New Deal period, when highway construction and improvement were undertaken extensively as relief measures, the emphasis was on secondary roads. Regular highway appropriations were actually reduced during the early 1930's, but the deficiency was more than made up by allocations of relief funds. The much-ridiculed WPA, for instance, spent $3.7 billion after 1935 on highway projects involving 600,000 miles of country roads and city streets.[20] In addition, in 1936 the federal-aid program was extended to secondary roads, used as Rural Free Delivery routes, farm-to-market roads, or school bus routes.[21]

While the United States was black-topping country roads, Germany was building its *autobahnen*, ostensibly for the same purpose of stimulating employment. The two situations were not, of course, identical. Nazi Germany did not have the same widespread ownership and use of automobiles as the United States, so that it had less need of a good secondary road network, and the *autobahnen* were planned with military use in mind. Each country was emphasizing the type of highway that it judged it needed most.

Yet for the United States this was not necessarily an "either-or" choice. The urgency of some of the local road improvement can be questioned. As late as 1950 the 350,000 miles of primary state highway, 11 per cent of all rural road mileage, carried 74 per cent of all rural traffic, and only 5.5 per cent of rural roads carried as many as 500 vehicles a day.[22] It seems regrettable that at a time when the cost of materials and labor was low, more was not done to create a network of up-to-date express highways. Without deprecating the substantial improvements made in American roads between the First and Second World Wars, the nation's highway policies accurately fitted the description given by Bernard de Voto in 1956: "In most places we are still in the curious condition of trying to adapt a road system designed for horse-drawn traffic to the necessities of the automobile age."[23]

The same statement can be made still more emphatically about towns and cities; indeed it has been made very recently by William L. Pereira, one of the leading contemporary American architects: "We are just now coming into the age of the automobile. And we have not built a single city as the result of the auto, because, as the case happens to be, it came too fast."[24] The emphasis is even stronger when we realize that Mr. Pereira was discussing Los Angeles, supposed to be par excellence the metropolis built about the automobile. Modern urban transportation is far too complex a matter to be considered in detail here; my purpose is simply to indicate some of the historical factors behind the present situation.

Until modern times, cities were always highly concentrated, partly for protection, but more because slow transportation made dispersal impractical. At the end of the nineteenth century commuter railroads and electric street-car and interurban lines permitted some outward expansion or suburbs, but rail-borne transport, because of its inflexibility and its need for high-density traffic if it was to operate efficiently, also tended to concentrate economic and recreational activities in the city center.[25] Street systems carried all kinds of traffic indiscriminately—through and local, carts, wagons, trolley cars, bicycles, pedestrians, and eventually automobiles—and the main thoroughfares, like the rail lines, converged on downtown. Consequently, when the motor vehicle began to appear on city streets in large numbers, congestion was inevitable—or rather, the congestion already existing was accentuated.

To make matters worse, when plans for trunk highways designed for automobile traffic began to be developed, the cities were simply left out. Belloc warned against this danger: "It is of little use to relieve traffic . . . between two urban centers if the exit and entry from and into each are blocked."[26] Yet even he proposed to carry his arterial roads only through the outskirts of cities, believing that traffic would disperse in the central area.

In practice, the *autobahnen* and the *autostrade* bypassed cities and to this day lack good connections with them, while in the United States the federal-aid program expressly excluded urban areas of more than 2,500 population. This policy was modified in 1934 to permit support for the urban segments of primary roads, but there was no provision for assisting cities to deal with their total traffic problem. Fundamentally the cities were left to fend for themselves, and few had the resources, even if they saw the need, to take such steps as constructing circumferential arteries so that through traffic could be separated from local and diverted from the crowded in-town area.

Instead, the main highways disgorged their traffic into every community along their routes, usually right through the middle, on streets already crowded with local traffic. To cite just one example, the motorist who took U.S. 40 from Philadelphia to St. Louis would not only follow part of the route of the Old National Road but would also become acquainted with the central business districts of Baltimore, Wheeling, Zanesville, Columbus, Indianapolis, Terre Haute, and Vandalia. When conditions became sufficiently intolerable, bypasses were improvised, generally inadequately marked, so that the through traveler avoided downtown traffic at the cost of getting lost in the suburbs.

Improvisation, in fact, describes most of the response of cities, American and other, to the advent of the automobile. To quote Wilfred Owen, "Most cities continue to settle for the unfortunate compromise of furnishing main highways to serve the dual purpose of moving traffic and providing access to land. These two functions cannot be supplied adequately by the same road."[27] Much that might have been done was proposed by William P. Eno before 1910: one-way traffic for narrow streets, painted lines and instructions on street surfaces, traffic signal towers, safety islands, and uniform regulations.[28] Unfortunately, these techniques, along with subsequent developments like timed traffic lights, were generally introduced as specific responses to immediate difficulties rather than as the result of systematic advance planning.

In addition to improved techniques of traffic control, there was some development of specialized roadways. Initially these were parkways, intended to separate pleasure from commercial traffic, or boulevards with a central roadway for through traffic and service roads for access to residences or businesses. The earliest of these appears to be Philadelphia's North East Boulevard, now Roosevelt

Boulevard, begun in 1903 and completed in 1918, with a 60-foot
central drive and 34-foot service roads.[29]

The preliminary steps toward New York's metropolitan parkway
system were taken before the First World War, but serious construction was not undertaken until the 1920's. The first parkways were not
built as divided highways, but they did have grade separation and
limitation of access. This example, however, was not widely followed.
This kind of construction was considered too expensive to be practical for intown areas. With few exceptions, parkways were restricted
to the outskirts of large metropolitan complexes. The Pulaski Skyway, built in this period between Jersey City and Newark, was a
further advance in that its planning introduced consideration of the
performance characteristics of the motor vehicle. Its designer, Sigvald
Johannesson, selected the high-level Skyway in preference to a low-
level bridge or a tunnel on the basis of a study showing that this was
the most economical solution in terms of cost of vehicle operation and
savings in time.[30]

The effective beginning of American highways designed for an
automotive age came in the 1930's. One of the first steps was the
extension of the New York parkway system into Connecticut by
means of the Merritt Parkway, which was completed by the end of
the decade and had a projected continuation, the Wilbur Cross Parkway. This was a limited-access toll road, with commercial vehicles
excluded.

More important in its long-term significance was the start of the
Pennsylvania Turnpike, undertaken as a project to relieve unemployment by converting the abandoned South Pennsylvania Railroad into
a superhighway. The railroad, planned in the 1880's to run from
Harrisburg to Pittsburgh, was never finished, and its graded right-of-
way, complete with nine tunnels, stretched across the state unused for
fifty years. The construction and operation of the road were put in the
hands of a Turnpike Authority, with power to issue bonds and
collect tolls.

Work commenced in 1937, with financial aid from the Public
Works Administration and the Reconstruction Finance Corporation,
and the initial section, 160 miles between the present Carlisle and
Irwin interchanges, was opened to traffic on October 1, 1940. It was
the first long-distance American highway designed for high-speed
motor vehicle traffic, both private and commercial: separate roadways, no crossing at grade, complete access control, no grade greater
than 3 per cent, and curves designed for speeds up to ninety miles an

hour. The major defect was that the Turnpike used seven of the nine railroad tunnels, and these accommodated only two lanes of traffic, so that they constituted a bottleneck in the system.

Further highway progress was halted by the Second World War, when shortages of labor and materials made possible only minimum maintenance on most of the nation's roads. There were exceptions; the Industrial Expressway between Detroit and Ann Arbor was built to serve the needs of war production, particularly the Willow Run bomber factory in Ypsilanti.

The war demonstrated the vital role that highway transportation had come to play in the life of the nation. When studies were made of the extent to which movement by road could be restricted in order to conserve gasoline and rubber, they revealed that 13 million people, almost a tenth of the population, lived in suburban communities that were not served by public transportation. The result was a fresh appraisal of highway needs, expressed in the Federal-Aid Highway Act of 1944, which provided for increased assistance to secondary roads, including their extensions into urban areas, and also for "a national system of Interstate Highways, not exceeding 40,000 miles in total extent, so located as to connect by routes as direct as practicable, the principal metropolitan areas, cities, and industrial centers, to serve the national defense, and to connect at suitable border points with routes of continental importance in the Dominion of Canada and the Republic of Mexico."[31] This program was based on highway research begun in the mid-1930's and embodied in reports to Congress on "Toll Roads and Free Roads" (1939) and "Interregional Highways" (1944).[32]

Congress authorized $1.5 billion to be spent on this total program in the three years after the war. It was not enough to get much work started on the Interstate network. The secondary roads eligible for aid totaled about 650,000 miles, and they absorbed much of the money. In addition, because federal support continued on the fifty-fifty matching basis, the states hesitated to assume their share of the heavy cost of superhighways. Nevertheless it was obvious to highway planners and engineers that in order to realize the very considerable potential of automotive transportation it was necessary to reactivate the principal formulated by John L. McAdam: "Roads must be built to accommodate the traffic, not the traffic regulated to preserve the roads."[33] And it was equally obvious what kind of roads these must be.

So, while the federal program lagged, the slack was taken up by two other programs under state sponsorship: the toll roads and the California freeways. Each applied the same technology of road construction (now officially designated "freeway"[34]) to quite different situations.

The toll roads were built principally in the Northeast, the conspicuous exceptions being Oklahoma and Kansas, for the primary purpose of expediting long-distance movement through regions of high traffic density. Charging tolls placed the financial burden on the traffic these roads were designed to benefit. Since it is possible for a passenger car to cross most northeastern states with, at best, one refilling of the fuel tank, state gasoline taxes would have borne most heavily on local populations, rather than on the out-of-state vehicles that compose, for example, more than half the traffic on the Pennsylvania Turnpike.[35]

The California freeways, on the other hand, were planned in the first instance to relieve congestion in the state's major metropolitan areas, Los Angeles and San Francisco. The great bulk of their traffic would be local, and since these two urban complexes accounted for the major proportion of all the state's automobile traffic, it made sense to finance the freeways by state taxes on motor fuels.

When the end of the Second World War permitted a resumption of major highway construction, the example set by Pennsylvania became the model for other states to follow. The first part of the Maine Turnpike, forty-seven miles from the New Hampshire border to Portland, was opened in 1947, designed specifically to relieve the appalling congestion that occurred every summer on U.S. 1 as it threaded its way through the resort communities on the Maine coast.[36] Then the turnpike network spread. The Pennsylvania Turnpike was extended across the state and continued by neighboring states to make a complete superhighway from New York to Chicago. The state of New York built its Thruway at a cost of a billion dollars and found it an excellent investment. Other major toll roads appeared in New England, Virginia and West Virginia, Kansas, and, of course, four turnpikes in Oklahoma.

The freeway system of California developed concurrently with the toll roads. The concept originated in traffic studies of Los Angeles during the 1930's, culminating with a report in 1939 recommending three hundred miles of freeways, later increased to six hundred, at an estimated cost of a billion dollars.[37] PWA funds permitted the building of the Pasadena Freeway and a short section through the Cahuenga Pass in 1940. After the war the Collier-Burns Act of 1947 gave California an up-to-date highway program, financed by gasoline taxes and providing state funds for metropolitan freeways. With this aid the Los Angeles plan could be implemented, and other California cities could follow this example—as, for instance, the construction of a freeway out of San Francisco to replace the highway that had become known as "Bloody Bayshore."[38]

There was still another significant highway project in this period. Route 128 in Massachusetts stands as a pioneering demonstration of the superhighway designed to carry traffic around the periphery of a metropolis, in this case Boston, and has been called a model for metropolitan bypass planning.[39] There was an earlier Route 128, a typical improvised bypass created by routing traffic through existing suburban streets. The concept of a circumferential freeway was first formulated in the 1930's, and a little construction was done, but the building of Route 128 as a complete planned highway was accomplished in the 1950's. Its effect was phenomenal. Even before its full seventy-mile length was finished, Route 128 was carrying 40,000 vehicles a day and an impressive collection of new industrial establishments had sprung up along it, valued at $133,500,000 by the end of 1957.[40]

The toll roads and the freeway systems of the 1940's and '50's not only performed an inestimable service in helping to handle the constantly increasing volume of motor vehicle traffic, but they also provided valuable lessons in the design and construction of such highways. Experience with the first turnpikes and freeways showed that either wider median strips or more effective divider fences were needed, that for long-distance travel some curves and other variations should be incorporated in order to prevent the phenomenon termed "highway hypnosis," and that on and off ramps required careful planning in order to permit smooth ingress and egress.

In addition, these roads gave impressive demonstrations of the gains in economy and safety that could be achieved with multi-lane, controlled-access, divided highways. The accompanying table shows the results of a study made by the Indiana Toll Road Commission of

Indiana Toll Road Truck Tests, Chicago–Jersey City

Factor	Turnpikes	Routes 30 and 22	Turnpike Savings
Elapsed Time	64 hrs. 49 min.	94 hrs. 43 min.	29 hrs. 54 min.
Travel Time	41 hrs. 5 min.	52 hrs. 22 min.	11 hrs. 17 min.
Gasoline Consumption	363.9 gals.	394.4 gals.	30.5 gals.
Speed Per Hour	40.93 miles	32.73 miles	8.20 miles
Gear Shifts	777	3,116	2,339
Brake Applications	194	890	696
Full Stops	58	243	185

Source: Indiana Toll Road Commission. These tests were made in April, 1957.

savings in truck travel between Chicago and Jersey City over the toll roads as compared with the parallel U.S. Routes 22 and 30.

The safety factor can be illustrated very simply with data for California. There the highway fatality rate for 1966 was 6.97 per 100 million vehicle miles on ordinary roads; 2.78 per 100 million vehicle miles on freeways. The freeways represent one-sixtieth of the state's highway mileage, but carry one-fourth of its traffic.[41]

These figures may even suggest that some of the safety agitation of recent years has been misdirected. Attention has been focused on the cars themselves, at the cost of ignoring the possible greater gains from properly engineered roads and traffic control systems.

The culmination of modern American highway policy was the passage of the Interstate Highway Act of 1956. This step was not the adoption of a new technology; it was the political and financial provision for the adoption of a known technology to create a coherent national highway system. It provided for 41,000 miles of express highways, including 5,000 miles of urban freeways, linking practically all urban areas of 50,000 or more population. The federal government pays 90 to 95 per cent of the cost of construction on a "pay-as-you-go" basis, through a Highway Trust Fund composed of a variety of highway-user taxes. The total cost is currently estimated at over $46 billion.

If this seems high, it is worth considering that the Interstate system, when complete, will comprise about 1 per cent of all the nation's mileage, but will carry 20 per cent of the traffic. The adoption of the Interstate program effectively stopped any further major construction of toll roads, since the states could now have their arterial highways built at federal expense. About 2,300 miles of the existing 3,000 miles of toll roads were incorporated into the Interstate system.

Interstate and urban freeways are not the whole story of highway policy. A complete account would have to consider what has happened on the other 3.5 million miles of roads and streets in the United States, from the primary federal-aid system to the 2.5 million miles that remain the responsibility of county and town authorities. There are also developing techniques of traffic control that deserve study, such as computer timing of traffic lights, metering access to crowded freeways, and adjustable dividers allowing lane flow to be reversed.

Nevertheless, in a society which has chosen to make extensive use of the motor vehicle, the main arterial routes ought to be designed as freeways. The Interstate Highway Act was an acceptance of this point—an acceptance long overdue.

The principles of controlled access, divided roadways, and grade separation were clearly understood by highway engineers in the 1930's, so that a network similar to the Interstates might have been started during this period of low costs. Such a policy would probably have had greater long-term value than the piecemeal improvement of secondary roads and streets that was actually done, and it would have been at least as effective in combating unemployment. The Interstate system could certainly have been effectively begun in the late 1940's, ten years earlier than it actually was, because all the basic planning had been done before the Interstate and Defense Highways Act of 1944 was passed. The long delay meant that construction had to be done in an era of constantly rising costs, with the result that completion of both the Interstates and of urban freeways fell behind schedule.

In summary, while we have had the motor vehicle in the United States since 1893, and the mass-produced automobile since 1913, it is only within the last twenty-five years that we have systematically planned a highway system designed for automotive transportation. Our highway policies have consistently lagged behind not only the needs of our traffic but also the highway technology that has been available to us. For this reason, highway programs have all too often fallen disappointingly short of what was hoped for them.

A Comparison of Rural and Urban Highway Mileage

(in 5-year intervals, in thousands of miles)

Year	Rural mileage	Urban mileage	Total mileage
1921	2,925	235	3,160
1925	3,006	240	3,246
1930	3,009	250	3,259
1935	3,032	278	3,310
1940	2,990	297	3,287
1945	3,012	307	3,319
1950	3,003	310	3,313
1955	3,057	361	3,418
1960	3,116	430	3,546
1965	3,183	507	3,690

Source: U.S. Department of Transportation, *Highway Statistics Summary to 1965*, p. 119.

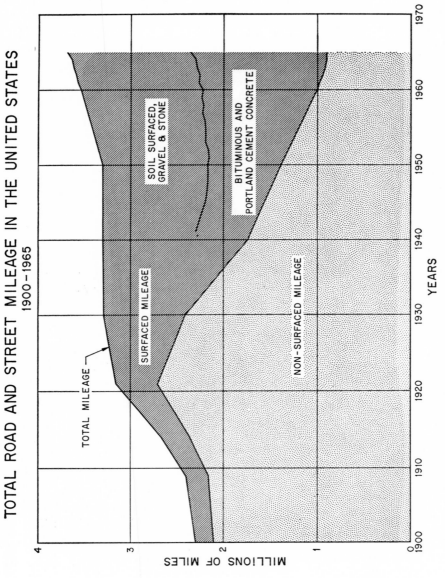

TOTAL ROAD AND STREET MILEAGE IN THE UNITED STATES
1900-1965

SOURCE: U.S. Department of Transportation, *Highway Statistics Summary to 1965*, p. 125.

1. Hilaire Belloc, *The Road* (London, 1924).

2. G. W. Taylor, *The Transportation Revolution* (New York, 1951), 132.

3. T. H. MacDonald, "History of the Development of Road Building in the United States," *Transactions of the American Society of Civil Engineers,* XCII (1928), 1194–95.

4. See P. W. Mason, *The League of American Wheelmen and the Good Roads Movement* (Ann Arbor, Mich., 1958).

5. MacDonald, "History of Road Building in the U. S.," *loc. cit.,* 1197.

6. *Automobiles of America* (Detroit, 1962), 103.

7. Charles L. Dearing, *American Highway Policy* (Washington, D.C., 1944), 81.

8. *U. S. Statutes at Large,* vol. 39, part 1, 355. The administration of the program was vested in the Bureau of Public Roads, the successor to the original Office of Road Inquiry. There were several changes of name. The agency became the Bureau of Public Roads in 1918.

9. F. L. Paxson, "The Highway Movement, 1916–1935," *American Historical Review,* LI, no. 2 (January, 1946), 243.

10. *U. S. Statutes at Large,* XLII, part 1, 212–19.

11. Paxson, "The Highway Movement," 246.

12. U. S. Department of Transportation, *Highway Statistics. Summary to 1965* (Washington, D.C., 1967), 167. State roads classified as primary but not included in the federal-aid program add another 150,000 miles.

13. *Ibid.,* 119.

14. Spencer Miller, Jr., "The Modern Highway in America," in Jean Labatut and Wheaton J. Lane, *Highways in Our National Life: A Symposium* (Princeton, N.J., 1950), 103.

15. *Ibid.,* 106.

16. *Ibid.,* 107.

17. Belloc, *The Road,* 196–99.

18. G. F. St. Clair and C. A. Steele, "Financing Highways in the United States," paper prepared for Fifth Pan American Highway Congress, Lima, Peru, 1944.

19. Paxson, "The Highway Movement," *loc. cit.,* 250.

20. C. L. Dearing and Wilfred Owen, *American Highway Transportation* (Washington, D.C., 1949), 107.

21. *U. S. Statutes at Large,* XLIX, 1519.

22. Miller, "The Modern Highway in America," Labatut and Lane, *Highways . . . ,* 97.

23. Bernard de Voto, "The American Road," in Ford Motor Co., *Freedom of the American Road* (Detroit, 1956), 8.

24. "Transportation: What's Ahead for Southern California?" Interview with William L. Pereira, *Westways,* Vol. 59, No. 11 (November, 1967), 5.

25. G. W. Hilton, "Rail Transit and the Pattern of Modern Cities: The California Case," *Traffic Quarterly,* Vol. 21, No. 3 (October, 1967), 379.

26. Belloc, *The Road,* 200.

27. Wilfred Owen, *The Metropolitan Transportation Problem* (Washington, D.C., 1956), 43.

28. "William Phelps Eno," *National Cyclopedia of American Biography,* LXVII (New York, 1965), 613.

29. Miller, "The Modern Highway," *loc. cit.,* 108.

30. *Ibid.*, 103–104.

31. *U.S. Statutes at Large*, LVIII, part 1, 838–43.

32. Department of Transportation, *1968 Highway Needs Report*, 90th Congress, 2nd Session, Committee Print (Washington, D.C., 1968), 4.

33. Quoted in C. L. Dearing, *American Highway Policy*, 20.

34. The Bureau of Public Roads distinguishes between "freeway" and "expressway"; the former has complete control of access and the latter partial control. This distinction, however, is subject to local usage. A number of roads with the title "expressway" are actually freeways by the BPR's definition. A toll road may be, and usually is, a freeway.

35. J. B. Rae, *The American Automobile* (Chicago, 1965), 188.

36. Charles Rawlings, "Maine's 'Big Rud,' " *Freedom of the American Road*, 34–36.

37. Paul T. McElhiney, *The Freeways of Metropolitan Los Angeles* (Los Angeles, 1959), 31–34.

38. Samuel W. Taylor, "Bayshore Campaign Aimed High," *Freedom of the American Road*, 52–54.

39. Frank Fogarty, "Boston's Magic Semicircle," *ibid.*, 42.

40. A. J. Bone and Martin Wohl, *Economic Impact Study of Route 128* (Cambridge, Mass., 1958), 13.

41. Davis Dutton, "Traffic Fatality Rates—a Downtrend for California," *Westways*, Vol. 59, No. 10 (October, 1967), 57. Comparable figures for the Interstate Highway system show a fatality rate per 100 million vehicle miles of 2.8 on Interstates and 9.7 on older trunk roads.

Commentary on the Paper of John Rae

By Carroll Pursell, University of California, Santa Barbara

SEVERAL YEARS AGO an outgoing public servant in California made something of a stir when he published a bit of muckraking titled "Confessions of a Highway Commissioner." In that essay he wrote, at one point, "I'm not an engineer, which may explain why I believe there are far fewer facts in this world than points of view."[1] I feel somewhat the same—I have no quarrel with any of the facts so ably presented by Professor Rae, but I would like to take issue with his point of view.

At the risk of caricaturing his treatise, I would restate it as follows: Since roughly the first decade of this century we have been in the Age of the Automobile, and stand in need of a highway system commensurate with this new age. The proper highway system is one made up of freeways specifically designed to move a large number of high-speed automobiles between large cities. Such a system was technically possible long before it was undertaken. The lag between need and response was owing to a lack of vision in general, and politics in particular.

I would suggest that several of these propositions need a closer look. I am struck, for example, by the problem posed by a California county official as early as 1855: "There should be," he complained, "two kinds of roads. Those connecting large cities and towns need to be located with greater care and with a view towards more permanency than the roads connecting villages. Although roads leading to distant localities may accommodate small villages this should be a secondary consideration. At present the reverse is the case. A zigzag line between villages has been adopted as a main thoroughfare and must be patiently meandered by travellers between two distant points."[2] Apparently the problem of properly separating through from local traffic was not a new one with the automobile.

Although Professor Rae is struck most forcibly by the late-coming of the freeway, I am equally struck by the fact that it has come to dominate so much of our thinking about transportation. At a meeting of the American Society of Civil Engineers in 1936 one member made an eloquent plea for what he called "the modern express highway," in fact a system of divided and limited access highways. Emphasizing the safety hazards presented by even the best highways of the time, he pointed out that several automobiles were then being advertised as capable of going over a hundred miles an hour. To which a colleague responded, "Is the fact that the manufacturer is providing something not needed and not in the public interest a valid reason for building roads to keep pace with it?"[3] It's a question we have answered only by default.

My own point of view would lead me to stand Professor Rae's question on its head: how has it come to pass that, the airplane aside, transportation has, in a few decades, come to mean almost exclusively automobiles, trucks, and highways? How is it, for example, that in the state of California some $800 million a year are spent on ever more costly and elaborate variations on an already baroque freeway system—and this at a time when there is a growing realization of the desperate need for public (and, we hope, rapid) transit? The answer, I suspect, provides us with yet another example of how attempts to substitute expertise for politics tend sometimes to emphasize the worst aspects of both. Once again California is a case in point.

According to the general law of 1850, new highways in the state were to be laid out by three disinterested persons appointed by the County Court. Six years later the Surveyor-General complained that "viewers are appointed who know nothing of the business they attempt and have no professional reputation to lose, and who wish the road to run where it will benefit them—instead of the public."

In an attempt to make road overseers more responsible, a new law in 1869 required that they be elected by the public they were expected to serve. It was a reform of dubious value, however, since the Road Fund became an attractive source of political patronage.

In 1883 the law was changed again to provide for the appointment of overseers by the county supervisors, and four years later it was again stipulated that they must be "disinterested" citizens to qualify for the post. Neither elected nor appointed officials at the county level seemed capable of coping with the state's road problems.

Under increasing pressure from such groups as bicycle enthusiasts and humane societies, the legislature in 1895 set up a State Bureau of

Roads which, like so many government bureaus, began as a modest research effort but eventually blossomed into one of the most powerful of all state agencies.

Technological innovations in road building and the increasing number of automobiles led to a shift from research and regulation to outright ownership and construction at the state level. Finally, as a result of the devastating earthquake of 1907, which destroyed a large number of bridges and ruined many main roads, the state set up a new Department of Engineering which included a division of highways. Soon three additional members were added to the Department to constitute a California Highway Commission charged with supervising the building of a state highway system.

The Division of Highways was modeled after a modern industrial enterprise, controlled by what John Kenneth Galbraith has recently termed the technostructure. A professional staff of engineers sought to turn away from the old tradition of amateur leadership, and to provide expert and disinterested administration of the highway system. Their dependence upon bond issues rather than annual appropriations helped give them a freedom of action not accorded most other bureaucrats, and when this method of financing began to prove inadequate in the mid-1920's, a Highway Fund was established from gasoline-tax monies.[4]

The whole structure was the very model of progressive government. An important technical problem was removed from the sordid control of mere politics and turned over to expert bureaucrats. Its organization was made businesslike in the most explicit meaning of that term. And perhaps most importantly, an independent source of income would presumably guarantee that roads would be located by considerations of public need and scientific fact rather than political influence. Engineers in the Federal Bureau of Reclamation had hoped for much the same kind of independence, but with less success.

The result was something at once more and less than the model of scientific method which engineers and progressive politicians had both hoped for. It is significant that the putative "Father of the California Freeway System" is not an engineer but Senator Randolph Collier of Yreka, chairman of the Senate Transportation Committee and leader of a highway lobby known as Randy's Rat Pack.

California today has a splendid system of freeways—a road system built specifically for the Automobile Age—primarily because highway engineers, encouraged by a powerful lobby of automobile clubs (which are mainly in the automobile insurance business), oil companies, trucking associations, heavy equipment manufacturers, and similarly interested parties, have virtually uncontrolled access to that $800 million a year which is earmarked exclusively for their use. The whole Freeway Establishment, as it has been called, is protected by an influential Senate committee which has thus far been able to prevent any raids on this transportation money for such socialistic proposals as those for public transport.[5] The civilian Highway Commission which is supposed to oversee the engineers is totally ineffective. "What actually exists," revealed an ex-commissioner, "is a condition wherein the inmates run the asylum. . . ."[6]

If this may indeed be characterized as the Automobile Age, it is not the result either of a Divine plan or of natural selection—but rather of a breakdown (or perhaps a subversion) of our political processes. It was not inevitable that our lawmakers should so serve the interests of the automobile or remake our cities in their image. It was, in fact, their responsibility to adopt a sound transportation policy which provides flexible and adequate service without sacrificing humane values. They have evaded this responsibility by turning the problem over to a group of men in whose professional and financial interest it is to subordinate all transport to the freeway. The result may be, as some progressive reformers hoped, a technocracy, but it is certainly not one from which we can take much hope.

The *New Republic* for January 25 of this year carried an article titled "Running Over the Public." It was stimulated by hearings held by the Federal Highway Administration to take testimony on whether or not federal law should require public hearings before new freeways are located. The American Association of State Highway Officials sounded the alarm and a solid rank of fifty states responded with dire warnings of the consequences of consulting the public on such purely technical matters. "What other state agency in the United States," marveled the *New Republic*, "could trick out a solid rank of governors to defend its bureaucratic autonomy?"[7]

Professor Rae began his essay with a quotation from Hilaire Belloc, and I would like to close my comments with something from the same source. Writing a poem for children, Belloc allowed himself a small dig at the technocrats.[8] He began:

> *The Microbe is so very small*
> *You cannot make him out at all,*
> *But many sanguine people hope*
> *To see him through a microscope.*

Belloc proceeded to describe this tiny beast—his jointed tongue, purple spots, and eyebrows of tender green, then closed with this injunction:

> *All these have never yet been seen—*
> *But scientists, who ought to know,*
> *Assure us that they must be so . . .*
> *Oh! let us never, never doubt*
> *What nobody is sure about!*

1. Joseph C. Houghteling, "Confessions of a Highway Commissioner," *Cry California*, I (Spring, 1966), 31.

2. Quoted in Gerald D. Nash, *State Government and Economic Development: A History of Administrative Policies in California, 1849–1933* (Berkeley, 1964), 121.

3. Discussion of Charles M. Noble, "The Modern Express Highway," *Transactions of the American Society of Civil Engineers*, CII (1937), 1102.

4. This sketch is taken from Nash, *op. cit., passim.*

5. Bob Simmons, "The Freeway Establishment," *Cry California*, III (Spring, 1968), 31–38.

6. Houghteling, *loc. cit.*, 29.

7. "Running over the Public," *The New Republic*, CLX (January 25, 1969), 12.

8. "The Microbe," H. Belloc, *Selected Cautionary Verses* (London, 1950), 120.

Commentary on the Paper of John Rae

By Eugene S. Ferguson, Iowa State University

A FEW DAYS BEFORE I received a copy of John Rae's paper I had just become acquainted with Leslie A. White's book, *The Science of Culture*, published in 1949. White argues for a deterministic kind of future, in which culture (meaning civilization) is the independent variable and man simply a necessary ingredient of culture.

One of the chapters in White's book is entitled "Man's Control over Civilization: An Anthropocentric Illusion." This translates roughly to: if we think we're in charge of anything, we're kidding ourselves. The author's argument on this point is, I must say, reasonably convincing. And when I read John Rae's paper in the light of White's book, I thought, "here is the perfect and complete confirmation of the theory of mindless technical development."

White's "culture" includes much besides technology, but I wondered also if maybe the development of automobile transportation doesn't suggest that technology alone is the determinant. That is, given the automobile and the road, all else will follow without more than an eighteenth-century Invisible Hand to guide the development toward the ultimate freeway pile-up, in which all the cars in California—not just a measly hundred or so—will be involved in a massive collision, and after which the debris will form a vast mound over downtown Los Angeles.

If I were to pursue that argument, however, I think we should quickly choose up sides, and with a subject as lively as highways we could expect to have a shouting match that would simply confirm my own point of view and have little effect upon *your* blind prejudices. So let me shift my ground a bit and ask the question: "what can I do with the information that is very clearly set out in this paper?" I am quite aware that John Rae may conclude that what I have done is twist his straightforward narrative into a horse for my particular

123

hobby. Nevertheless, I'll proceed to the question: "what can I do with 'The Car and the Road'?"

The most obvious answer, to me, is to use it as text material in a college course in the history of technology. But as I read it over I found the same difficulty that I have with much, if not most, of the writing that has been done for the classroom (and this includes my own writing). The objection was stated several years ago by Charles Gillispie in a way that has stuck with me. In a review of volume 3 of Singer's monumental *A History of Technology*, he observed that the historian of technology has a problem different in kind from that of the historian of science. ("At least," he wrote, "we [in the history of science] *have* a discipline.") He threw out a suggestion or two as to how the history of technology might be organized, and then he closed his review with the sentence that confirmed my own experience. He wrote, "I do not . . . know what I could find in this volume to teach to a course of undergraduates."[1]

I'd like to take a minute to say why I find it difficult to use this kind of text material. I have engineering students, by and large, in my classes. They are there in order to "take a course in" a non-technical subject that appears to them to be less unpalatable than other non-technical subjects they have heard about. They know—and their technical instructors confirm their knowledge—that the only important courses are in their vocational subjects, but a certain minimum number of non-technical courses are required by accrediting agencies.

Because of their vocational commitment, my students have almost certainly accepted without reservation the technological society in which they have grown up. Their faith is in technology, and only a few of them are aware that questions are being asked about where our technical virtuosity is taking us, except perhaps the questions by cranks and college professors who have been misled by the Communists. When these students read the quantitatively impressive technical record of automobile and highway building over the last fifty years, their reaction is, quite predictably, "so what else is new?" They let me know in explicit fashion that this certainly has nothing to do with them, because they are interested in the future, not the past. All the standard arguments get me nowhere.

Charles Gillispie would have us give more generalizations in the history of technology. I'm inclined to agree, but I find that any generalization that I can think of on the basis of this paper will be either pretty flat or unsupportable. A flat generalization might be: "highway building will continue at an accelerating rate in the foreseeable future." An unsupportable one might be this: "fewer problems would occur in the future if engineers were given full authority to determine highway policy."

Now, I don't expect to solve the problems of historiography in a brief commentary, but I do think we ought to recognize historiography as a central issue if the history of technology is ever to have any bite for students who desperately need to be bitten.

Let me suggest, therefore, some generalizations of a different order that might grow out of a study of American highways. The point of view of the study would have to change, many more data would be required, and I am not sure that these data would be admitted by the committed and the faithful. At the very least, however, I think we'd have something to argue about.

Here is my first generalization: "Notions of saving money through the use of more sophisticated technology are always mistaken." I think that the statement of 1916 in the "Car and the Road," that "neither freight nor passengers will ever be carried long distances over roads as cheaply as they could be over railways," is still largely true. We ought to be more attentive to the distinctions between cheapness and preference. We frequently confuse "cheaper" with "less effort" or "more fun," as, for example, in most computer applications. We might, for instance, test an assumption that is made by engineers, economists, and the Pentagon. This is the assumption that a cost-benefit study (a balancing of costs against expected benefits) is *not* simply an exercise in making legitimate the gleam in the eye of a prophet of progress, of justifying a decision that has already been made on some grounds other than economic. To say that new highways are justified by cost-benefit studies is simply to say that our economic studies are so arranged that they will confirm our prior decision to build new highways.

A second generalization: "Successful entrepreneurs always make sure that public subsidies, especially those that are unintended, are exploited fully." Long distance trucking happened to the United States almost inadvertently. After the First World War had shown the possibilities, a few individual hustlers built a network of long-distance truck lines. Eventually, fleets of trucks were competing successfully with the railroads.

The Association of American Railroads, during the 1930's and '40s, published a number of pamphlets pointing out the rather obvious subsidies that truckers enjoyed. The truck associations, when they were asked to testify before Congressional committees, were able to turn their critics aside with arguments of personal service and convenience.[2]

In the 1970's, with the completion of the new federal interstate highway system, free-enterprising truckers will fill the roads with three- and four-unit trucks, running at seventy miles per hour, just as soon as state legislatures can be brought into line by an aggressive and effective truck lobby. The Ford Motor Company has been for at least three years promoting its gas-turbine driven, twin-trailer truck (developed first with Army then Navy contracts), designed specifically for the interstate highway system; and Caterpillar Tractor Company is ecstatically pushing a three-trailer truck which is already being operated by Western-Gillette, a trucking firm. "For many months now," says the Caterpillar magazine, "this aggressive trucking firm has been hauling triples in Nevada . . . where regulations have encouraged full-scale tests."[3] The public interest, as I see it, is well on its way to being effectively circumvented.

My third and last generalization (I have Carroll Pursell, Jr., to thank for this one,[4] and although I have thought of two or three others, my time is running out): "While every major technical innovation sets loose unexpected forces, engineers always act as though the effects of their work are both limited and predictable." The three-lane road, which was designed by an engineer, even if he didn't think of the idea, ought to put us on our guard against leaving any significant decisions to engineers. Even on a four-lane freeway in Iowa, two head-on collisions were required to convince engineers that motorists were not just being perverse when they took the wrong turn and found themselves headed into a stream of traffic approaching at sixty miles per hour.

Finally, I seem to remember reading, in the 1950's, flat denials by automotive engineers that automobiles were serious offenders in the growing smog of Los Angeles. The current enthusiasm for steam automobiles, if it continues along present lines, will have unlooked-for effects, such as a new set of exhaust products and because of inherently low thermal efficiency a radical increase of exhaust heat, which is no longer negligible in cities. No serious discussion of the long-range problems of steam automobiles, nor indeed of any alternative power plant, has yet been started.

In closing, let me suggest again that there are alternative conclusions that may be, and I think should be, reached through the study of the history of technology. Because automobiles, highways, and nearly every major technical innovation have had unexpected and undesirable effects as well as expected and desirable effects, it seems evident to me that a radical change ought to be made in the way we think about our technology.

I cannot hope that the tax-free status of the Society for the History of Technology will be endangered by any campaigns that may be mounted to influence legislation, but I do think we ought to see if we can't find more in the history of technology than a series of success stories flecked only by failure of the moneyed interests to let technologists run the show. If we can put before students a past they'll argue with, rather than simply ignore, I think we might find ourselves with a significant as well as lively discipline.

1. *Isis*, L (June, 1959), 165.

2. Representative pamphlets are: Association of American Railroads, *Highway Motor Transportation* (Aug. 1945), and American Trucking Associations, Inc., *The Case for the Trucking Industry* (June–July, 1950).

3. Ford Motor Co., *Down the Road*, a pamphlet distributed to groups that listen to the description, almost entirely technical, of the new trucks. I heard the Ford presentation in 1965. *Caterpillar Energy Management*, II, No. 1 (1968), 5.

4. Carroll W. Pursell, Jr., ed., *Readings in Technology and American Life* (New York: Oxford, 1969), 248.

Art, Technology, and Science:
Notes on Their Historical Interaction

By Cyril Stanley Smith, Massachusetts Institute of Technology

IT IS MISLEADING to divide human actions into "art," "science," or "technology," for the artist has something of the scientist in him, and the engineer of both, and the very meaning of these terms varies with time so that analysis can easily degenerate into semantics. Nevertheless, one man may be mainly motivated by a desire to promote utility, while others may seek intellectual understanding or aesthetic experience. The study of interplay among these is not only interesting but is necessary for suggesting routes out of our present social confusion.

Humanists have shown a widespread disregard for technology's role in human affairs, but if they had seen technology as an eminently human experience, they could have better guided society's choice of objectives and controls. Civilization has been an ecological process with interacting contributions coming from an infinite diversity of individual human characteristics and social institutions. As historians have turned away from their older concern with the great movements headed by kings, generals, or businessmen, they have naturally emphasized the role of people like themselves (scientists and other intellectuals), and they have, until recently, largely disregarded the rather messy technology that has been associated with virtually every important historical change and which continually impinges directly upon Everyman in his daily life. Neither religious conviction nor institutional conservatism has, until today, sensed in technology a peril sufficient to prompt an examination of its nature and its growth. Certainly, at the extremes, the concepts of the cosmos and of the ultimate nature of matter developed by philosophers and scientists are of overriding importance, for they have basically influenced man's opinion of himself: Men have gone to the stake for their ideas on the nature of the universe, and all men know of it. Ideas on ultimate atomism have aroused bitter philosophic debate. Conversely, however, anyone

who considers the nature of materials, advocates a new way of making pottery, or advances a new theory of the hardening of steel meets
with both intellectual and popular indifference. Yet the voyage to the
moon depends on men making metal as well as on computations
based on the theories of Newton and Einstein.

The present paper is an outcome of my realization, some years
ago, that many of the primary sources I had selected for a study of the
history of metallurgy were objects in art museums. Though materials
are not all of technology, they have been intimately related to man's
activities throughout all of history and much of prehistory and therefore provide an excellent basis for a study of some of man's most
interesting characteristics under greatly different social and cultural
conditions. A materials-oriented view of history may overemphasize
the association of technology with art; yet it was precisely the artist's
search for a continued diversity of materials that gave this branch of
technology its early start and continued liveliness despite an inner
complexity which precluded scientific scrutiny until very recently.

Art, Techniques, and Materials

Several writers have discussed the manifest interactions between
artistic expression and the basic view of the world embodied in contemporary scientific or religious concepts. Such interactions certainly
exist at the highest level of insight, but artists have had far more intimate and continuing association with technology than they have had
with science. In turn, the attitudes, needs, and achievements of artists
have provided a continuing stimulus to technological discovery and,
via technology, have served to bring to a reluctant scientific attention
many aspects of the complex structure and nature of matter that simplistic science would have liked to ignore. The antecedents of today's
flourishing solid-state physics lie in the decorative arts. One must conclude that creative discovery in any field is a matter for the whole
man, not his intellect alone. Though it occurs in an individual mind, it
is strongly interactive with society and tends to seek out the least rigid
parts of a community structure.

Leonardo da Vinci said in his treatise on painting: "Those who are
in love with practice without science are like a sailor who gets into a
ship without rudder or compass and who never can be certain where
he is going."[1] At the same time, Leonardo strongly opposed the view
that knowledge that is both born and consummated in the mind is
enough: "It seems to me that all sciences are vague and full of errors

that are not born of experience . . . , that do not at their origin, middle or end pass through any of the five senses."[2] And, of course, all his extant works reflect continual interplay between sensual experience and intellectual analysis. The same view is to be found in the writings of many scientists, though for most of the last three centuries science has rightly been more concerned with the unreliability of the senses than with their essential contribution to whatever knowledge human beings can acquire.

When discussing the new routes to the understanding of nature in the preface to his *Micrographia* (1665), Robert Hooke remarks:[3] "So many are the links upon which the true Philosophy depends, of which, if any one be loose, or weak, the whole chain is in danger of being dissolv'd; it is to begin with the Hands and Eyes, and to proceed on through the Memory, to be continued by the Reason; nor is it to stop there, but to come about to the Hands and Eyes again, and so, by a continual passage round from one Faculty to another, it is to be maintained in life and strength." Hooke believed that the advancement of knowledge depended upon both senses and the intellect— upon the mind, the hand, and the eye in cooperation. His writings repeatedly reflect his obvious enjoyment of natural phenomena and his intuitive understanding of them. However, Hooke's slightly younger contemporary Isaac Newton was engaged in demonstrating the great power of mathematical science and setting the stage for three centuries of superbly unfolding knowledge based on the belief that the senses are unreliable and that science advances best if, at any one time, it is limited to those small areas in which rigorous methods can be applied. Though the domain accessible to such science is steadily expanding, there remain many important aspects of natural and man-made systems that are too complicated for complete analysis. The present-day political and intellectual unrest reflects increasing awareness that the scientist's understanding of things "in principle" is not enough. The more holistic view of the Renaissance artist may be returning—though whether it will be put into practice by people who allow themselves to be called artists is another question.

Just as the meanings of the words "art," "science," and "technology" have varied greatly throughout history, so has the role in society of the various practitioners. Perhaps technology has been the most constant in its aims. Science has encompassed many different approaches to the collection and analysis of data, just as art, in different places and periods, has combined in vastly different degrees the functions of decoration, symbolism, illustration for didactic purposes, the projection of feeling, and (by no means the least important) pure enjoyment. In what follows the "art" may sometimes be of a kind beneath the notice of an art historian, but it will always be concerned with a man's doing something that is not strictly necessary for the performance of a function, something extra done to give enjoyment to the producer himself and usually also to others who subsequently come in contact with his work.

Not all peoples have regarded "art" as a separable human activity, and the self-conscious production of paintings, sculpture, and *objets d'art*, like the organized commerce in them, has by no means always occupied the privileged place that it has had in Europe since the Renaissance. Most of what follows is concerned with the decorative arts—those arts relegated to the minor category in most museums today—although it might be remarked that the best of today's non-objective paintings have more in common with sensitively wrought useful objects of ceramic and metal than they have with many of the "fine" arts displayed on museum walls.

There is some analogy between the exploration and exploitation of the materials of nature in chalcolithic times and earlier, the detailed exploration of the forms of nature that followed increased representational skill in the thirteenth and fourteenth centuries, and the experiments with perspective, light, and shadow in the Italian Renaissance. The driving force in all three was an essentially scientific curiosity directed to the discovery of some fairly practical means of achieving an aesthetic end.

The relation between art and the artist's materials was well discussed by Henri Focillon.[4] Remarking that art is bound to weight, density, light, and color, he says that it is borne along by the very matter it has sworn to repudiate: matter in its raw state "evokes, suggests and propagates other forms according to its own laws." The ceramics of the Far East appear "to be less the work of a potter than a marvellous conglomerate created by subterranean fire or accident. The raw stuff of [Chinese ink brush painting] partakes of both water and smoke . . . yet . . . such a painting possesses the extraordinary secret of being able to stabilize these elements and at the same time leave them fluid and imponderable." Though overemphasis on technique is clearly dangerous, Focillon believed that "the observation of technical phenomena not only guarantees a certain objectivity to [the studies of] a historian but affords an entrance in the very heart of the problem by presenting it in the same terms and from the same point of view as it is presented to the artist." In discussing the artist's various techniques to get different qualities of line, shadings, and graduations, "such alchemy does not, as is commonly supposed, merely develop the stereotyped form of an inner vision: it constructs the vision itself; gives it body and enlarges its perspectives."

Technique, of course, mainly gives details of form, not the gross outlines and balances. Nevertheless, much of the refinement of an artist's vision as he works toward its realization comes from his inter-action with his materials. The whole quality of a line and surface depends upon both the material and the tool as well as upon the ✓ artist's hand, whose movements they subtly control. Compare the same pictorial concept as it is realized in different media—with a brush in oils, watercolor, or tempera on canvas, wood, or paper; by printing from a metal plate with intaglio lines made by etching or engraving or from surfaces left in relief on a chiseled wood block; or by repoussé work, tracing or otherwise working directly on the final metal surface. It is understandable that those students who must work from reproductions of works of art are usually more interested in iconography than in questions of technique and quality, but it is re-grettable that technical ignorance should so frequently prevent art historians from considering the whole experience of the artist. In much the same way, science historians have tended to overlook the less logical side of science.

The Discovery of the Properties of Matter

In studying man's earliest history, when the evocative qualities of certain forms and the power of symbolism in nonrandom shapes and sounds was being discovered, it is difficult to separate things done for "pure" aesthetic enjoyment from those done for some real or imag-ined "practical" purpose. The man who selected for admiration a beautifully shaped and textured stone was yielding to a purely aesthetic motivation, but the man who molded clay into a fertility figurine was simultaneously an artist, a scientist learning to under-stand the properties of matter, and a technologist using these proper-ties to achieve a definite purpose. Supposedly most of the innumerable fertility figures recovered by the archaeologist's spade from periods even before 20,000 B.C. were made as a kind of industry, acquired for reasons of fashion, and employed practically to make more probable some desired result. This does not, however, destroy their funda-mental aesthetic quality.

More important is the fact that in the earlier stage of discovery, first of form and later of materials that, once shaped, would retain desir-able form, the motive can hardly have been other than simply curiosity, a desire to discover some of the properties of matter for the purpose of internal satisfaction. Paradoxically man's capacity for aesthetic enjoyment may have been his most practical characteristic, for it is at the root of his discovery of the world about him, and it makes him want to live. It may even have made man himself, for, to elaborate a remark by the poet Nabokov, it seems likely that verbal language (to which anthropologists now assign vast evolutionary advantage) was simply a refined use of the form-appreciating capabilities first made manifest in singing and dancing.

A natural step after the collection and admiration of unusual natural stones and animal or vegetable debris would have been the use of the properties of some natural materials to produce unnatural shapes and textures in others. This supposedly began by matching the hard cutting edge of stone to softer wood, hide, sinew, and bone, and was followed by the discovery and exploitation of the special properties of a host of substances. The last were mainly minerals that could be ground and used as pigments, undoubtedly far more for the decoration of the body and other long-perished surfaces than for the incredibly preserved cave paintings that we admire so greatly today. It is not only the nature of the record that makes one feel the joy that early man took in the discovery of the properties of materials. The cracking propensity of different stones, the plasticity of moist clay, the fine granular color of pigments were all used for what they are and appreciated directly by the senses in shaping or in use.

Aesthetically satisfactory forms have repeatedly developed from interaction between cultural requirements and the real properties of a new material or technique: the forms are not just superimposed. A returning sensitivity to this is at least partially behind the present passion for primitive art, for a simpler technology makes the properties of materials more evident.

Over and over again scientifically important properties of matter and technologically important ways of making and using them have been discovered or developed in an environment which suggests the dominance of aesthetic motivation. The presence of flowers in Neanderthal graves[5] suggests that the transplanting of flowers for enjoyment preceded the development of agricultural technology for food supply. The first use of both ceramics and metals occurs in decorative objects. Fire-hardened figurines of clay precede fired pots in many Middle Eastern archaeological sites. The seventh millennium B.C. copper dress ornaments and beads at Chatal Hüyük in Anatolia and at Ali Kosh in Iran considerably precede the use of copper for weapons, though the useful needle appears early. Although there is

Fig. 1   Gold beaker and cups made by raising from sheet metal, decorated by repoussé work and tracing. Cup height 15.5 cm. From the Royal Graves at Ur, ca. 2600 B.C. (*Photograph courtesy University Museum, Philadelphia*)

Fig. 2   Gold rings made from square and round wire almost invisibly soldered. 2.1 and 1.7 cm. diameter. From the Royal Graves at Ur, ca. 2600 B.C. (*Photograph courtesy University Museum, Philadelphia*)

Fig. 3 *Left*, Chinese cast bronze ceremonial vessel, type *tsun*. Early Shang Dynasty. Height 27 cm. (*Photograph courtesy Arthur M. Sackler Collection, New York*)

Fig. 4 *Right*, Chinese cast bronze ceremonial vessel, type *ting*. Late Shang Dynasty. (*Courtesy Fogg Art Museum, Harvard University*)

Fig. 5 *Below*, Chinese cast bronze bell. (detail). Chou Dynasty. The design is built up in three successive stages of replication. (*Courtesy Freer Gallery*)

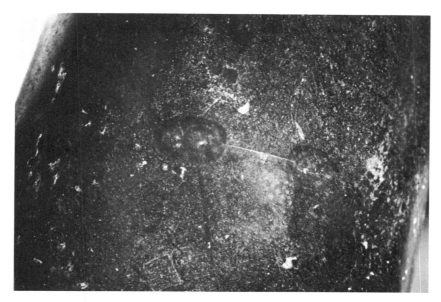

Fig. 6 Welded joint in Roman cast bronze statue, natural size. (*Courtesy Arthur Steinberg and the Museum of Fine Arts, Boston*)

Fig. 7 *Below*, Section of cast-on leg in late Chou Dynasty bronze vessel. (*Photograph by John Gettens, courtesy Freer Gallery*)

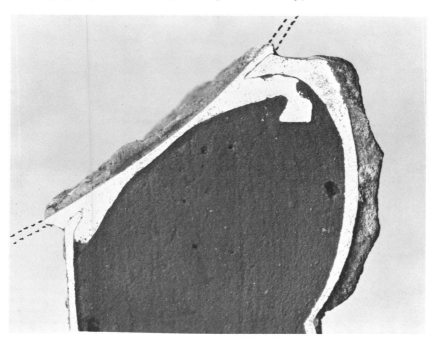

Fig. 8 Blast lamps being used in making colored enameled glass beads. Johan Kunckel *Ars Vitraria experimentalis* (1679). (*Photograph courtesy Corning Museum of Glass*)

Fig. 9 *Below*, Cast bronze funerary bucket with lathe-tool marks on bottom. Roman, ca. 200 A.D. 25 cm. diameter (*Courtesy W. J. Young, Museum of Fine Arts, Boston*)

 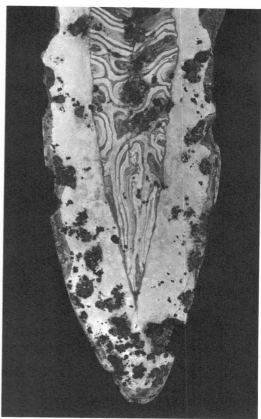

Fig. 10 *Left*, Ivory cup. One of the products of the 19th century ornamental turning lathe, showing some of the complicated shapes made possible by the mechanical combination of simple motions. J. J. Holtzapffel, *Turning and Mechanical Manipulation*, Vol. V, London, 1884.

Fig. 11 *Right*, Tip of a pattern-welded iron sword. Merovingian (6th century A.D.). Width, 3.9 cm.

Fig. 12   Etched design on Italian helmet, Milan, 16th century. (Detail.) (*Courtesy John Woodman Higgins Armory, Worcester, Mass.*)

Fig. 13   "Instantaneous light box" with case made of green *moiré métallique.* Made in London about 1820 by "J. Watts and Co., Chymists No. 478 Strand." Height, 8.0 cm. This device made fire by bringing a wooden match tipped with potassium chlorate and sulphur into contact with concentrated sulphuric acid. *Moiré métallique* was tin-plated iron that was given a special treatment to develop a fancy crystallization, subsequently etched and covered with colored lacquer. (*Photograph courtesy Bryant and May Ltd. and Science Museum, London*)

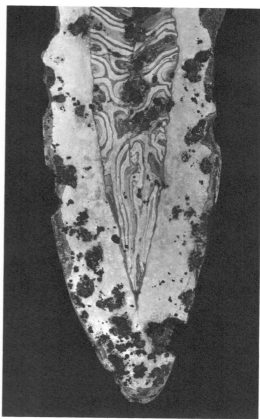

Fig. 10 *Left*, Ivory cup. One of the products of the 19th century ornamental turning lathe, showing some of the complicated shapes made possible by the mechanical combination of simple motions. J. J. Holtzapffel, *Turning and Mechanical Manipulation*, Vol. V, London, 1884.

Fig. 11 *Right*, Tip of a pattern-welded iron sword. Merovingian (6th century A.D.). Width, 3.9 cm.

Fig. 12   Etched design on Italian helmet, Milan, 16th century. (Detail.) (*Courtesy John Woodman Higgins Armory, Worcester, Mass.*)

Fig. 13   "Instantaneous light box" with case made of green *moiré métallique*. Made in London about 1820 by "J. Watts and Co., Chymists No. 478 Strand." Height, 8.0 cm. This device made fire by bringing a wooden match tipped with potassium chlorate and sulphur into contact with concentrated sulphuric acid. *Moiré métallique* was tin-plated iron that was given a special treatment to develop a fancy crystallization, subsequently etched and covered with colored lacquer. (*Photograph courtesy Bryant and May Ltd. and Science Museum, London*)

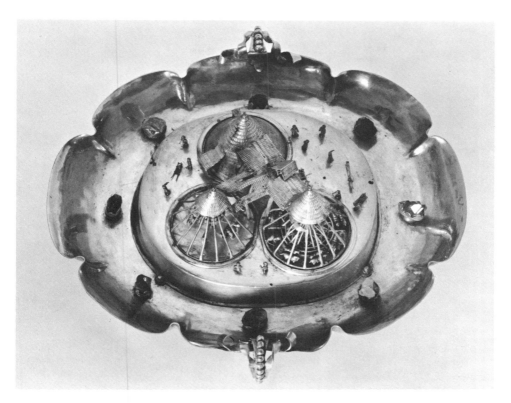

Fig. 14 Copper dish (Herrengrund ware) containing model of minehead equipment with working miners and mineral specimens. Length, 33 cm. Heavily gilded. The inscription. *"Eisen war ich, Kupfer bin ich. Silber trag ich, Gold bedecket mich,"* refers to the recovery of the copper from mine waters by displacement with scrap iron. Made in Herrengrund, Bohemia, early 18th century. (*Photograph courtesy Abegg-Stiftung, Bern*)

Fig. 15 *Below*, Print from one of the earliest American electrotype plates. This formed the frontispiece of Daniel Davis, Jr., *Manual of Magnetism* (Boston, 1842), alongside a print from the original engraved plate, indistinguishable from the copy.

Fig. 16   A door knocker—the first use of malleable cast iron. From R.A.F. de Réaumur *L'art de convertir le fer forgé en acier . . .* (Paris, 1722), plate 16.

Fig. 17   Design for a key of forged and chiseled iron. From M. Jousse, *La Fidelle ouverture de l'art de serrurier* (1627), plate 1.

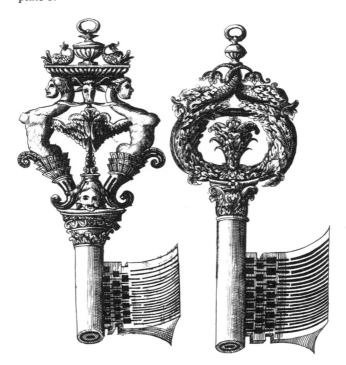

Fig. 18   Woodcut view of an assay laboratory. From Lazarus Ercker *Beschreibung allerfurnemisten mineralischen Ertzt und Berckwercksarten* (Prague, 1574).

Fig. 19   *Below*, Woodcut diagram to illustrate the bending stresses in a beam. Galileo, *Discorci e dimostrazioni matematichia intorno a due nuove scienze* (Leiden, 1638).

Fig. 20 Engraving showing "nothing else but the appearance of a small white spot of hairy mould, multitudes of which I found to bespeck and whiten over the red covers of a small book." Robert Hooke, *Micrographia* [1665], plate 12. The scale line is 1/32 inch, corresponding to an original magnification of about fifty.

Fig. 21 *Below*, Engraved diagrams showing paths of rays of light in the eye and in other media. R. Hooke, *Micrographia* (1665), plate 6.

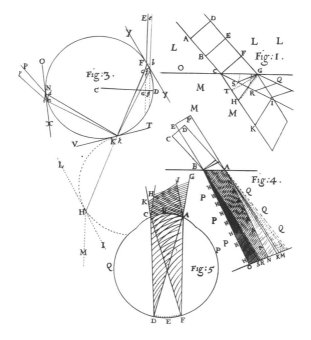

Fig. 22 Engraving showing machinery for blanking and striking coins.
André Félibien, *Principes de l'Architecture* (Paris, 1676).

Fig. 23 Interior view of workshops producing hammered copper vessels. From Duhamel du Monceau, *"Déscription de la manufacture du cuivre de M. Raffaneau établie près d'Essone,"* appended to −. Galon, *L'Art de convertir le cuivre rouge . . . en laiton . . . .* ([Paris], 1764).

Fig. 24 *Below,* Inlaid mosaic decoration on columns at the palace at Uruk, ca. 3500 B.C. (*National Museum, Berlin. Photograph courtesy Bildarchiv Foto-Marburg*)

Fig. 25 Gold earring in polyhedral form
composed of gold granules accurately soldered together and unconsciously illustrating the concept of the crystal lattice. From
Marlik, ca. 1000 B.C. The granules in the
top and bottom tetrahedra are of different
sizes and their junction in the central plane
illustrates an intercrystalline boundary. (*Photograph courtesy Iran Bastan, Tehran*)

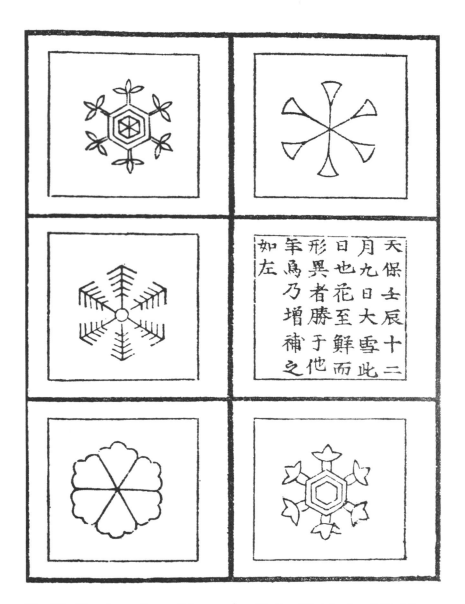

Fig. 26 Drawings of snowflakes made with the "Dutch glass." Toshitsura Doi, *Sekka zusetsu* (Tokyo, 1833).

Fig. 27   Japanese sword guard with snowflake design. Iron with inlay. Goto School, ca. 1850. (*Photograph courtesy Toledo Museum of Art*)

Fig. 28   *Below*, Mosaic tile work in the Alhambra at Granada, ca. 1325 A.D. (*Photograph by Phylis Morrison*)

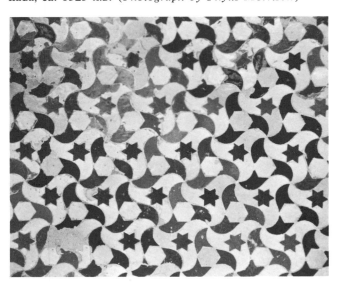

Fig. 29 *Ballet magnétique* by Greek-American sculptor Takis (1968). (The solenoid, intermittently energized, causes irregular movements of the two pendulum bobs, which are suspended on steel wires that strike the taut transverse wire, producing musical notes and dancing of the suspended rod at the left.)

some evidence for earlier pyrotechnological experiments with ores, the replacement of simply hammered native copper by smelted metal did not occur until about the time that copper oxide was being used in blue glazes on ceramics, though probably only after high temperatures had become available for firing useful ceramic sickle blades.[6]

The modern metallurgist uses alloying elements to strengthen metals and to lower their melting point; he cold-works to harden and anneals to soften. He uses their differing chemical reactivities, immiscibilities, and surface energies in refining and joining processes. The discovery of all these effects is very old. To take a single point in history, an examination of the jewelry and other metal objects from the famous royal graves at Ur,[7] dated about 2600 B.C. (Figs. 1 and 2), reveals knowledge of virtually every type of metallurgical phenomenon except the hardening of steel that was exploited by technologists in the entire period up to the end of the nineteenth century A.D. One must not, of course, overlook the fact that royal burial objects are far from being representative samples of contemporary use of any material. The court would appropriate the best work to its own ends, but just for this reason it provides the best index of both the most novel and the most sophisticated techniques.

The transition from copper ornaments to axes and swords of bronze in the fourth millennium B.C. was paralleled in the fifteenth and sixteenth centuries A.D. by the transition from the casting of monumental bronze doors, statuary, and especially bells, to the casting of cannon. If the objects themselves are not sufficient evidence, a comparison of the vivid circumstantial account of bell founding given by the early twelfth-century artist-craftsman Theophilus[8] with the discussion of the casting of cannon by the eminently practical Biringuccio[9] some four centuries later will show how much the warrior depended upon the churchmen's technique. To be sure, existing technology is applied to whatever need may be seen by a government or people: my point is only that the *invention* of a technique has, until recently, been more likely to occur in an aesthetically sensitive environment than in a practical one. We will see later that even the development of efficient quantity-production methods owed much to the art industries, if not directly to the artist.

Technology's debt to the artist is inseparable from the converse. Though both the most exquisite and the most ugly objects can be made with the same technique, technique is essential to beauty. The technique of the artist merges by invisible stages into the technology of his materials. Among the best examples of this are superb cast bronzes of Shang Dynasty China.[10] The earliest ones reflect an advanced ceramic technology, so necessary in making the molds, and have linear decoration obviously cut into the mold (Fig. 3). The almost brutal strength that characterizes the later bronzes of the Shang and early Chou periods, the flanges, and almost every aspect of their form arise in a direct interplay between design and the practical details of the foundry. The molds were divided into a number of sections that would have produced unpleasant breaks in the surface decor had these not been designed for division, and leaky mold joints would have produced ugly fins if these had not been exaggerated into flanges the edges of which could easily be dressed (Fig. 4). The attractive difference in quality of the fine and the bold intaglio lines probably arises in the technical difference between carving the former directly into the mold surface and making the latter by applying convex lines of clay to a molded concavity. Still later comes the building up of designs from a few units by the use of some method of three-dimensional replication (Fig. 5)—a clear forerunner of the printing process.

In view of the centrally important role that welding plays in today's space-age structures, it is interesting to note the facility with which Greek and Roman founders welded together the parts of their statuary. Almost any classical bronzes, when closely studied, reveal some patching of foundry defects, but recent studies[11] have uncovered the widespread use of some process (not yet fully understood but clearly involving the running-in of superheated molten metal) for making joins between precast parts, with an accuracy and permanence that would challenge a modern welder (Fig. 6).

The technique of casting-on preceded this joining of long seams and was widely exploited in the Middle East, in Europe,[12] and in the Far East[13] because of the freedom that it gave to the designer (Fig. 7). It also enabled different metals to be combined, as in the application of an elaborately cast bronze handle to a serviceable steel blade—a literal welding of beauty and utility—and in the combination of cold-worked and cast bronze.

In later times the little mouth blowpipe used by the goldsmith in his soldering operations was adapted to laboratory use, first for the examination of ores and later as the basis of the first comprehensive scheme of qualitative chemical analysis.[14] The larger blast lamps worked with foot bellows that were used for making glass beads and for decorating them with colored enamels (Fig. 8) led directly to the oxy-gas blowpipe. This was first used in high-temperature research about 1782, became fully commercial with the melting of platinum in the 1850's, and finally became the modern welding torch.

The decoration of pottery with colored pigments and later with glazes repeatedly brought to man's attention the chemical diversity of natural minerals and led to new techniques. The cementation process that was probably used in the fourth millennium B.C. to make Egyptian blue frit (faience) involves very subtle behavior of alkalies and silicates in differential contact with lime and silica surfaces.[15] It is highly probable that it gave rise directly to the manufacture of the first "sandcored" glass vessels (which probably had a calcareous, not a siliceous, core). Though its relationship to early metallurgy has not yet been explored, this cementation process gives hints of the way in which the first alloys may have been made. It may relate to the smelting of complex sulphide ores by the use of highly alkaline fluxes, to say nothing of its later use in the soldering, parting, and coloring of gold and eventually in the making of brass and steel.

In the eighteenth century, European desire to duplicate beautiful porcelain from the Orient inspired not only geological search but also experiments in high-temperature chemistry and the development of the first realistic methods of chemical analysis for anything but the precious metals. Reports of the large-scale operations at Ching-te-Chen may have inspired the integration of mass-production operations at Wedgwood's factory in Staffordshire, and, at the other end of the spectrum of knowledge, it was an interest in porcelain that led to Réaumur's studies of the devitrification of glass, which later played a role in the understanding of lava and the development of Hutton's plutonic theory of the earth.[16]

Chinese fireworks for pleasurable celebration inspired more diverse chemical experimentation than did military explosives. Today's rocket ships and missiles are an outgrowth of fun-fireworks, and their guidance systems depend on knowledge first acquired from that ubiquitous toy, the top. All optical devices have their roots in the polishing of ancient mirrors and the cutting of accurate facets on gems for a more decorative glitter. The chemist's borax-bead test, now alas passé, arose from the use of metal oxides in making stained glass windows and colored enamels (as well as fake gem stones).

Colors and chemistry are inseparable. The earliest pigments were naturally occurring minerals, but the preparation of artificial ones, such as red and white lead, verdigris, and marvelous sublimed vermilion, mark a chemical industry in classical times. The subtleties of surface tension on which the modern flotation process for the beneficiation of ores depends were first used in the purification of lapis lazuli to give fine ultramarine.[17] The important metal powder industry of today began with gold ink.[18] Art historians rarely go behind the

blue and gold splendor and the iconography of a medieval illumi-
nated manuscript to see the ingenious technology that made it possible and that reflects men's lives on another, no less necessary, level.
It is the same with organic dyes: think of the chemical knowledge
behind an Oriental rug or an emperor's robes! The chemist's indi-
cators and his eventual awareness of pH came directly from the
chameleon colors of the miniature painter's turnsole. . . . The list
is endless.

The Development of Mechanical Technology

The relation between design, structural engineering, and knowledge of materials in architecture is a well-known example of the
inseparability of aesthetic and technological factors. Here it must
suffice to make only the passing comment that it has usually been
nonutilitarian structures such as temples and monuments that have
stretched the limits of existing techniques and led to the development
of new ones.[19]

The popular belief that technology is recent is partly based on the
fact that intricate machines were, in fact, slow to develop. The advanced knowledge of materials in the ancient world was not paralleled
by mechanical devices of seemingly comparable ease of discovery.
The ancient military devices (which have usually followed not far
behind aesthetic needs in promoting discovery) and hoisting machines of importance to the builder are all relatively simple. Mechanical devices of any intricacy appear only as toys, as aids to priestly
deceptions, or as theatrical machinery. It was not utility in the usual
sense—though it may have been a search for the public's money—
that prompted the mildly ingenious devices described by Hero of
Alexandria.[20] It may be, as has often been suggested, that the avail-
ability of cheap labor rendered the Persians, Greeks, and Romans
unable to appreciate the advantages of mechanical power; but their
failure to develop other types of intricate mechanisms is, I believe,
attributable to the fact that the aesthetic rewards to beginning experimentation by the curious in this area are not large. Indeed, for
simple mechanical experiments to be intriguing, they require a kind
of overlay of intellectual analysis: they are too easily reproducible to
provide a rich and varied sensual experience of the kind that comes
directly from play with minerals, fire, and colors. Not until the mid-
twentieth century have artists shown much desire to experiment with

machinery, and their efforts sometimes seem to be more directed toward catching up with and exploiting the technologists' world than toward leading it.

The association of the earliest clocks with mechanical automata was a natural one, for, with the possible exception of organ makers, only the makers of automata had the necessary skill and sense of mechanism.[21]

Machine tools, like materials and mechanisms, had a period of prehistory within the decorative arts. The earliest is probably the rotary drill, which, though it was perhaps developed for hafting axes, found wider use in making beads, seals, stone pots, and sculpture.[22] The inverse geometric motion of material against a fixed tool begins with the potter's wheel and progresses to the simple lathes that supposedly produced the soft-stone products of Glastonbury and the Roman bronze objects such as mirrors and pots having decorative bottoms with deep, heat-catching circular grooves (Fig. 9). Then followed Theophilus's twelfth-century description of lathes for turning bell molds as well as for the molds of pewter pots and the metal pots themselves. By this time rotary motion was commonly used in the grindstone. The first machine with intermittent motion after the Oriental rice-pounding mill is Theophilus's little device for cutting the criss-cross ground for decorative overlay of precious metal on iron. The cam- and template-guided lathes of Jacques Besson (1578) not only cut screws but also turned decorative woodwork of great variety. They were followed by the ornamental turning lathes of the seventeenth to nineteenth centuries, used mainly by gentlemen hobbyists and for decorating gold snuff boxes. These were devices of great mechanical ingenuity applied to a mechanically trivial purpose (Fig. 10); nevertheless, they provided the experimental environment in which definable motions were generated not only as a basis for instrument making and later industrial machinery but also to disseminate a feeling for the composition of mathematical curves.[23] The toy-like nature of these lathes resulted in their being rather briefly dismissed in the standard machine-tool histories,[24] but it is easy to see how the desire to produce a decorative effect was once more the motivation for the discovery of phenomena that would later be applied to more serious purposes.

Decorative fountains—for example, Versailles with its magnificent pumps and pipes—stretched the capacities of hydraulic engineers more than did plebian water supply. Savery's fire engine was pumping water for a garden in Kensington in 1712 not long after its use in mine drainage. The even more portentous principle of Jacquard's loom (1801) was not needed for weaving plain fustian but for the fanciest of lace.

The introduction of printing illustrates the same point, though here the art is even less separable from the technology. The obvious advantage of transmitting information in written form kept thousands of scribes busy for millennia, but the functional business of recording the commands of the government or the information needed by merchants did not lead to printing—this came from the desire to reproduce images and patterns. The ceramic decorative stamps at Chatal Hüyük,[25] the cylinder seals made in such profusion throughout the Middle East, the tools for the impression of decorative details in ceramic vessels and tiles as well as in molds for casting, the punches for repetitive stamping of metal, the dies for striking metal coins, and the block printing of textiles—all these precede "useful" typographic printing and lay the groundwork for it. The sequence from rubbing to woodblock print to movable type in the Far East is a direct one.[26] The first true printing was for the dissemination of a Buddhist sutras —utility and aesthetics united in the service of religion. In Europe, although the precise stages of the invention are hard to trace, the sequence is similar. The reproduction of pictures with text from woodblocks was a popular art early in the fifteenth century, though for the step to reusable type Gutenberg's solution involved the transfer of technique from a humbler craft, that of the pewterer, whose permanent molds with replaceable parts for decorative detail and whose alloy needed little change to make type.[27] The earliest type seems to have been cast from a tin-base alloy perhaps containing bismuth, but cheaper, harder lead alloys were common in the sixteenth century and thereafter.

A strong aesthetic motivation is visible in the works of the early typographers. Much of it obviously derived from the desire, or perhaps the necessity, of duplicating the quality of the manuscripts with which they were initially in competition. But art and technology are even more inextricably interwoven in the reproduction of pictures, which began before typography but received an enormous impulse from their use to illustrate printed books. Though to some extent the mere possibility of making multiple copies is the enemy of art, limited reproduction brings an artist's works to a greater audience, and the techniques themselves give rise to aesthetic qualities not otherwise obtainable. Woodcuts, etchings, lithographs—especially if the artist's hand prepares the printing surface—are often preferable to unique works executed in the traditional media of the painter.

Print-making from intaglio lines in metal plates was late in appearing, but its roots are deep. Decorative engraving on the surfaces of bone or soft stone objects, of course, precedes the use of metal, and it was widely used pictorially on three-dimensional objects of bronze, gold, and silver. The earliest date on a print from an engraved plate is 1446.

Some playing cards printed about four years later have attractive animal designs that are similar to some of the marginalia in the great manuscript Bible of Mainz (dated 1452–53, now in the Library of Congress), and Lehmann-Haupt[28] has suggested that the plates may have originated in abortive experiments by an engraver working in collaboration with Gutenberg, who at the very time and in the same city was at work on his famous Bible and would naturally have liked marginal embellishment matching the best contemporary manuscripts to appear alongside his typographic text. Plausible and attractive though this hypothesis is, there is no intaglio printing that can be definitely associated with Gutenberg.

In any case, for hints as to possible technical steps behind the invention, we must move to Italy, where the first engraved prints—those of Maso di Finiguerra, 1452–55—were made slightly later than in northern Europe. Sulphur casts associated with the Italian prints are preserved in both the British Museum and the Louvre.[29] Goldsmiths were accustomed to make such replicas of engraved objects, both to check the designs before filling them with niello and to provide a record for future use. It was a simple matter to make a mold (perhaps of plaster) from the engraving and to obtain an exact replica of the original intaglio lines by casting sulphur in it; smearing this with soot and oil would make the design clear and produce a general effect of black lines on a yellow background much like the final niello on gold. Transfer to paper would follow naturally and soon render the cast copies obsolete. Northern engravers may have been more ingenious: the casting of the mold material on a dirty engraving might have suggested direct transfer to paper without the need for double molding or a sulphur intermediary. In any case, fine prints could not have been made in the fifteenth century had not centuries of earlier work with niello developed both the technique of using the graver as well as the sense of design appropriate to it, had not the caster of art bronze had experience in the replication of models with fine detail, and had not the new oil-based inks and presses become available for the printer.

Once the process of transferring to paper had been invented, it spread rapidly, and artists throughout Europe produced prints which used to the full the possibilities of rendering fine detail and controlled shading that were implicit in the technique. A few years later, shortly after 1500, engraved plates began to meet competition from those in which some or all of the lines were bitten with acid, giving them a special quality that many artists prefer to engraved ones. Here, too, an old technique was ready to be adapted to a new purpose, for armor makers had been using etching in the decoration of the more elaborate of their products for at least a century, probably much longer.[30]

The beginnings of etching—the removal of metal by localized chemical attack—are obscure. Supposedly jewellers and copper-smiths had long used vegetable acids or minerals such as misy to re-move the oxide scale produced by annealing their ware, but there is no early record of this practice. Chemical attack was certainly used in the cementation process to remove silver from solid gold at least as early as the sixth century B.C.[31] The earliest etching done with a re-served design is actually that on calcareous shells at Snaketown Pueblo in Arizona,[32] dating from the first century B.C. A related effect of a slightly later date is seen in the reserved areas of silver in the depletion-gilded sheet ornaments of copper-silver-gold alloy found in northern Peru.[33]

In the Old World, swords made of different kinds of iron and steel welded together into a consciously decorative pattern appear in La Tène sites, and supposedly some chemical attack would have been used to reveal the pattern. Quite apart from its decorative function, the visibility of the pattern in the welded composite would serve simultaneously to control the work in the smithy and to provide a kind of index of quality to the customer. The pattern on the swords of the Franks and Vikings (Fig. 11) are referred to in Viking sagas in terms that leave no doubt as to their visibility.[34] The beautiful textures of Damascus swords were also acclaimed by poets long be-fore the technique of forging and etching them was described.[35] These were certainly etched to bring out the pattern, and etching was probably done on the European blades, though polishing alone can leave a just-visible texture on the surface if there is enough slag mixed in the metal. Japanese swords owe both their effectiveness and their beauty to the distribution of intensely hard areas left by an intricate control of the forging and heat-treating operations. These, with finely dispersed slag particles, are subtly revealed in the final polishing operation. There is, in fact, no better symbiosis of the highest aesthetic and technical standards[36] than in these swords.

Some paragraphs in Pliny may refer to chemical attack on iron, but the first clear reference to etching in European literature is in the eighth-century chemical manuscript at Lucca, *Compositiones variae*, which contains a recipe for the treatment of an iron surface with a mixture of corrosive salts containing copper as a preliminary to gilding it. A similar technique appears in the ninth-century *Mappae clavicula*.[37] With the omission of copper and the use of a stop-off to localize the effect, decorative etching was born (Fig. 12). Although the earliest extant etched decoration is on late fifteenth-century iron armor, there is earlier evidence for its use. Conrad Kyeser's 1405 manuscript, *Bellifortis*,[38] describes the preparation of distilled nitric acid for this purpose, and he even calls it *aqua martis*, in clear reference to its use on iron. It seems highly probable that the discovery of this first mineral acid about a century earlier had come directly from the experimental distillation of an etching mixture containing saltpeter and acid sulphates. Parenthetically, hydrochloric acid, distilled from a mixture of chlorides and sulphates, also appears first in connection with decorative embellishment—in a work on dyeing, *Plictho*, published in 1548—and in 1589 decorative etching with it is described, but on marble, not metal.[39]

The technique of etching passed directly from arms to the production of etched iron plates for printing, which was at first a part-time activity of armorers. But, having begun as art, etching eventually began also to influence science. As the Damascus and Merovingian swords showed, etching is a sensitive means of revealing heterogeneity in steel, but metallurgists did not begin to use it consciously for this purpose until 1762. In the period between 1773 and 1786, observations on the etching of Damascus gun barrels, which were then being made in Europe, led to the first identification of carbon as the material responsible for the differences between wrought iron, steel, and cast iron.[40] The investigation of an essentially decorative phenomenon, and an oriental one at that, thus led directly to the most important single scientific discovery in metallurgical history!

Soon thereafter etching gave rise to a new decorative technique known as *moiré métallique*.[41] This was invented in 1814 and aroused considerable excitement for a few decades. (Fig. 13 shows a fire lighter made by this technique.) It was simply etched and lacquered tin plate, but the plate was sometimes treated by local heating and cooling to give very fancy crystallization patterns, even semblances of flowers and landscapes!

New methods of printing illustrated books repeatedly redounded to the advantage of both science and technology. An interesting printing technique—first described in the sixteenth century[42]—was to make direct impressions of objects such as leaves by coating them with printer's ink and impressing them directly on paper. The process (which is not unrelated to the much earlier and more versatile oriental method of producing rubbings on paper laid over objects with details in relief) was later called "nature self-printing." In the eighteenth century a number of botanical books were published with illustrations printed this way, the first being J. H. Kniphof's *Botanica in originali*, published in 1733.[43] The same technique was used by von Schreibers and Widmanstätten in 1813 for recording the etched structure of a section of the Elbogen meteorite. Their print was a spectacular improvement in clarity and accuracy over the lithographs of other meteorites that accompanied it in their published book[44] or the engravings by Gillet de Laumont in the *Annales des Mines* of 1815. For a time thereafter many methods of obtaining relief or intaglio impressions of an object directly on a printing surface were experimented with for both scientific and other purposes.[45] Nature printing from a collage of textured surfaces is the basis of a flourishing school of printmakers today, and books illustrated by nature printing are being issued by private presses in England.

The early history of photography itself is a classic example of the symbiosis of art and invention. Della Porta in 1558 recommends the *camera obscura* as a device to lighten artists' labors and help them with perspective. Niepce's famed photochemical etchings on glass (1826) were done to reproduce art, not reality. The processes of Daguerre and Talbot were of both worlds, as photography has been ever since. When the invention of photomechanical methods displaced most other methods in the printer's shop, etching had become a common laboratory technique. The science of metallography— indeed, practically the whole structural side of modern materials science—stems from the work of Henry Clifton Sorby in 1863–64 in the famous steelmaking center of Sheffield, which was the world center of supply for engravers' steel plates. By applying to the preparation of laboratory specimens the methods used for giving these plates their fine finish, and by using etching, which he had heard discussed at a meeting of the local Literary and Philosophical Society, Sorby was able to reveal for the first time in history the true microstructure of steel without disfigurement by fracture or deformation.[46] In the present connection it is interesting to note that the next paper on metallography, by the German railroad engineer Adolf Martens in 1878, was directly inspired by some work on the quality control of metal for use in the exquisite art castings of iron for which Germany was rightly famous at the time.

Electroplating and Electrical Engineering

Electrochemistry is another area in which the interest of the artist or the art industry accelerated scientific knowledge and technological development. An old and pretty parlor trick was the *Arbor Dianae*, mentioned with other "metallick vegetations" in most chemical textbooks of the seventeenth and eighteenth centuries. Eighteenth-century assayers knew of the electrochemical series (though they did not call it that) in the form of sequential replacement of silver in solution by copper, copper by iron, and iron by zinc.

The medieval use of an acid cupriferous solution to give a coating of copper on iron was mentioned above. Such electrolytic replacement remained a common observation and was sometimes used for recovering copper from waste mine waters as well as to confuse people with the semblance of transmutation. It gave rise to a minor art in the seventeenth century in the form of a very pleasant ware made from cement copper in the town of Herrengrund in the Bohemian Erzgebirge.[47] These objects bear inscriptions reflecting their polymetallic origin (as in Fig. 14) or cryptic jingles such as on a wine cup in my possession, which reveals a common source of scrap iron for the process:

> *Ein Pferd mich vor mit füssen trat,*
> *da ich noch Eisen ware,*
> *durch ziment wassers baad*
> *bring ich gut freünd zu baare.*

It has been reported that this ware was shaped in iron and then plated by immersion in cement water. Some folk objects were certainly so made, but the real Herrengrund ware is actually nonmagnetic and was probably made from cement copper powder that was melted, refined, cast, hammered into sheets, and shaped as any other copper would have been. The role of electricity in these operations was not, of course, suspected, anymore than it was in the mysterious decay of rudder irons on the English ship *Phoenix* that in 1670 had been sheathed with sheet lead, which had just then become available in wide sheets from the new rolling mill at Deptford.[48]

If any of these effects had been looked at by a sufficiently curious mind, Galvani's discovery could easily have been made a century or more earlier and without the intervention of a frog. However, even after Galvani and Volta, even after Wollaston's and Cruickshank's demonstrations of the cathodic deposition of copper and other metals, and even after Michael Faraday's elucidation of the laws of electrolysis, no use was made of the phenomenon until 1838. In that year Jacobi, Spencer, and Palmer, in somewhat confused priority, all began the art of electrodeposition for the duplication of coins and other small art objects[49] as well as for the reproduction of printing surfaces—at first for illustrations (Fig. 15) and later for letterpress.

Henry Bessemer later claimed that in 1832, when a young man of nineteen, he had reproduced plaques by electrolysis, but he did not publish. After 1838 this quickly became a very popular hobby and resulted in widespread knowledge of electricity. Smee, writing in 1842, remarked that "there is not a town in England that I have happened to visit, and scarcely a street of this metropolis [London] where prepared plasters are not exposed to view for the purpose of alluring persons to follow the delightful recreation afforded by the practice of electrometallurgy." The new metallurgy quickly spread from copper to other metals. Gold was naturally one of the first, but the most commercially significant was the electrodeposition of silver upon the beautiful white copper alloy now known as nickel silver which rapidly displaced the more expensive Sheffield plate. The base alloy itself had been imported from China for about two centuries and was used for fireplace equipment, candlesticks, and other domestic objects of some elegance,[50] but it was not analyzed until 1776 (twenty-five years after the discovery of nickel itself), and it was not available commercially until 1833, just in time for its wedding with the new plating method.

Electroplating soon became an important art industry. It provided economic support for the beginning of the nickel industry, it provided a large market for electromechanical generators before the development of electric light, and, with telegraphy, it provided training and experience for innumerable men who were soon to combine their empirical knowledge with a growing science to give birth to electrical engineering.

Art for the Masses

The story of electroplating is only one example among many in which a desire to simulate a precious material in a cheaper form has stimulated technical advance. All students of the historical literature on iron and steel know Réaumur's classic *L'art de convertir le fer forgé en acier et l'art d'adoucir le fer fondu* (Paris, 1722), with its curious subtitle *faire des ouvrages de fer fondu aussi finis que de fer forgé*. Réaumur specifically states that his incentive was to provide cheaply for the masses decorative objects of the kind expensively made and previously available only to the rich. It was not engineering devices but elaborately chiseled door-knockers (Fig. 16) at which he first aimed—but he ended by extolling the virtues of mass production of interchangeable parts in industry generally.[51] The motive of cheap art was also behind Réaumur's development of his "porcelain," a devitrified glass of the type that has recently been revived in superior form.[52] He was also the first to suggest the use of wood pulp in the making of paper.

Today's steel rails, I-beams, and other structural shapes also have their origin in decorative needs—the rolling of H-shaped lead cames for stained glass windows. Around 1750, fancily profiled sections of iron for use in balcony railings, window moldings, and the like were being made in grooved rolls three decades before Henry Cort applied the process to the large-scale consolidation of wrought iron bars.[53]

Parcel gilding can be justified on purely aesthetic grounds as producing an agreeable contrast in color, as in inlay, but most gilding operations have been done simply to save money and to make expensive-looking objects available to others than the rich. The preparation of thin gold leaf, the most extreme utilization of the malleability of any metal, is similarly inspired. The fact that composition gradients could be produced in solid metals was made quite clear, long before diffusion became a subject of scientific inquiry, by the common use of gilding via gold amalgam, as in Europe, or by chemical methods of surface enrichment, as in pre-Columbian South America and in Japan.

The many changes of properties and surface coloration of metals produced by goldsmiths could hardly have failed to support the belief that transmutation is possible—as indeed it is, if "transmutation" is not limited to nuclear changes but is applied to major changes of physical properties.[54] Hopkins,[55] in particular, has argued that alchemy was an outgrowth of the joining of Greek philosophy with a knowledge of workshop practices. Yet the value of empirical knowledge naturally fades as a field advances, and the replacement of alchemy by modern chemical theory is attributable more to the logical than to the practical approach.

The above examples show that the art industries have contributed greatly to the development of techniques and to the knowledge of reactions on which today's science and technology are based. Perhaps, indeed, the mixture of aesthetic and commercial motivation involved in such developments was quantitatively the most powerful stimulus of all, for basic discovery of new effects inspired only by curiosity is by its very nature rare, as rare, indeed, as any great individual work of art.

Little reliance can be placed on any of the written sources relating to technological history prior to about A.D. 1500 unless they are confirmed by contemporary nonverbal evidence. Even today technologists are not noted for literacy, and men like the Benedictine monk Theophilus (early twelfth century), whose hands were accustomed to both the hammer and the pen, have always been rare. In books ideas naturally fare better than technology. Moreover, the chances of survival of written technological information in medieval libraries were not high. For all this, there are many records that must be studied in the absence of anything better, and in these a strong bias toward the decorative arts is evident. Following the Roman Vitruvius's *De architectura*, the best pre-Renaissance technological sources are the Leyden papyri, the Lucca manuscript entitled *Compositiones variae*, the *Mappae clavicula*, Theophilus's *De diversis artibus*, and Eraclius's *De coloribus et artibus romanorum*. Every one of these deals with the artist's materials and techniques, to the exclusion of almost everything else. Manuscripts presenting primary information on machinery for warfare, mining, and other industrial occupations do not exist before the fourteenth century. The famed Theophilus's manuscript of about A.D. 1123 is an outstanding source of pure technology, though he confines himself to giving intimate details on painting, stained glass, and metal work for the embellishment of the church.

Although it was far from the artist's conscious intent, many paintings of religious subjects, especially those of the thirteenth through the sixteenth centuries, convey information on agricultural and building techniques, and they are particularly important in reflecting current attitudes toward labor and machinery. Lynn White, Jr.,[56] has studied from this viewpoint the changing depiction of the seven Virtues. As late as the twelfth century Temperance was in little esteem, but during the thirteenth century she became identified with measure and subtly associated with internal and external control. By 1350 she is depicted with the newly invented sandglass; by 1450 (in a manuscript now in Rouen) all the seven Virtues are depicted with technological appurtenances, but Temperance displays eyeglasses, rowell spurs, a mechanical clock, and a tower windmill. The showing in such a scene of these objects—all recent inventions—expresses "a reverence for advancing technology, a sense of its spiritual value, which is peculiar to the West and which has been essential for the building of industrial society."

There are innumerable illustrations known to historians of art but only little used by their technical confreres in which an artist interested in human activity (either for its own sake or to satisfy an ecclesiastical or princely command) used the decorative aspects of tools and mechanical devices, quite commonly with disregard of mechanical details but nevertheless providing useful information to the historian.[57] And, of course, the very materials of the artist are themselves a superb record of the technology that produced them, a record that can be read in intimate detail by modern laboratory techniques. The output of the spectroscope, microscope, and X-ray spectrometer will soon become as important to the technological historian as his older verbal sources, to which we now return.

The two earliest printed works on the prosaic subject of steel both have artistic overtones. The first was a little pamphlet on etching, the *Stahel und Eysen* (1532) mentioned above, and the second was on ornamental ironwork and locks—Mathurin Jousse, *La fidelle ouverture de l'art de serrurier*, published at La Flèche in 1627 (Fig. 17). By this time the artist was aiding the technologist in substantial ways, for the new techniques of producing more accurate representation of visual appearances served increasingly to convey precise technical information. The accurate, detailed drawings of the liquation process for the desilverization of copper, and those of lathe details, pile drivers, etc., contained in the fifteenth-century Nuremberg "Hausbuch"[58] and of the various crafts in the Hausbuch of the Mendel Brothers' Foundation[59] are a far cry from earlier illustrations in which technology is only incidentally reflected. One of the leading German illustrators of the sixteenth century, Jost Amman, sought inspiration in the technical crafts for eighty-six of the 118 woodcuts in his popular *Book of Trades*.[60]

By the mid-sixteenth century many carefully written books on both science and technology were being printed with woodcut illustrations (Fig. 18). Both the biological sciences and technology required and inspired some of the best efforts of the artist in rendering realistic details without confusion. The woodcuts in the well-known treatises of Vesalius, Agricola, and Ercker are about as attractive as any book illustrations of the sixteenth century, and their instructional value was correspondingly high. The mystical side of alchemy, though scientifically sterile, appealed to the artist's imaginative approach and gave rise to some attractive books.[61]

In general, physics, with its abstract concepts and simple diagrams, neither attracted nor needed the artist. If a physicist used illustration at all, they were likely to be in the form of colorless linear diagrams making visible the geometry implied by his equations. Galileo in 1638 depicts a weed-encrusted stone wall supporting his elastically-deflected beam (Fig. 19), but later elasticians eschewed such realism. The terrellae in William Gilbert's *De magnete* (London, 1600) have a pleasant look, perhaps contributed by the man who cut the block, and to illustrate his observations on the magnetization of cooling iron, Gilbert allowed himself the luxury of including a woodcut view of a blacksmith's shop that is in the direct tradition of the series of such views in the medieval *Speculum humanae salvationis*, where they illustrate (amid changing hearth and anvil design and with occasional detachability of the horse's leg to simplify the smith's work) metallurgy's first contribution to the fine arts—Tubal Cain's rhythmic clangor giving rise to the idea of melody in a listener's mind. Gilbert's other illustrations, however, are purely linear diagrams.

The engravings in Robert Hooke's *Micrographia* (1665) reflect both approaches. Most of these are well-shaded sensitive representations of exciting vistas in the New Landscape that his microscope was exposing for the first time (Fig. 20), but Hooke's diagrams of the paths of rays of light (Fig. 21) have a sharp austerity which matches the abstraction of the idea and which came to characterize most scientific diagrams thereafter.

The engraved copper plates preferred by seventeenth- and eighteenth-century book publishers over wood blocks permitted accurate delineation of apparatus and were excellent for showing machinery (Fig. 22). The enormous growth of the graphic arts in eighteenth-century France coincided with a rationalist viewpoint to result in the publication of massive collections of engravings of technical subjects. The hundreds of folio-size plates in the series of *Descriptions des arts et métiers* published by the French Academy of Sciences (Fig. 23) and the seven volumes of plates accompanying Diderot's famed *Encyclopédie* provide a profuse record of technical crafts and industry. Our knowledge of the technology of that time is probably more complete than that of any other period in history, for before this there was scant interest in making records and after it the profusion of technology both outran the possibility of fully recording it and stifled an interest in the details of its minor variations.

On the Segregation of Disciplines

The conscious separation and classification of an activity or viewpoint as science, technology, or art is recent and came about rather slowly. It is misleading to apply modern classifications to earlier periods in which distinguishable professions did not exist and a desired end result dominated over conscious particularities of method. Nevertheless, it is obvious from the above that I regard the somewhat less fully intellectualized activities of the technologist as having much in common with those of the artist and, until recently, interacting rather less with those of the scientist.

The Renaissance marks a natural interaction between a rejuvenated art and a beginning science.[62] In the fourteenth century many artists delighted in using their newly awakened powers of observation and their increased skill in representation to embellish the margins of manuscripts with precisely limned naturalistic living forms,[63] while at the same time they came to observe and emphasize the essential aspects of their subjects in a manner that later became appropriate for scientific illustration. Conscious studies of the interaction of light with matter and almost mathematical considerations of perspective are reflected in the mid-fifteenth-century paintings and sculpture by Ghiberti, Brunelleschi, and others, who both set the tone of the new times and absorbed its spirit. Fifteenth-century writing about art is very different in tone and intent from the earlier collections of pigment recipes or the practical how-to-do-it treatise of Theophilus. Stillman Drake[64] shows how the conflict of theory and experiment in sixteenth-century music contributed directly to the development of the style of Galileo and other great seventeenth-century scientists. Yet, as the different viewpoints that had been combined in the artists' activities came to be consciously realized, an inevitable result was that each of them should become a separate field of specialization.

The artist became an important individual, highly visible in society, while craftsmanship took on a markedly lower status. Logical thought had always aided the artist in making his materials conform to his vision, but when the critical interplay between logic and experiment was consciously separated as a method of learning about the world, it became the new science, growing and changing beyond all recognition of its origins. For four centuries now it has outrun the other aspects of the artists' approach and has done so by exploiting the power of partial isolation. If mathematics could deal with music and perspective, it could also deal with falling bodies—but it could handle the planets better than a terrestrial feather, for it only applies to ideal isolated systems of simplified forces and bodies, one or two or at most a very few things at a time. Science in its very essence is simple. The new physics could deal with ideally elastic bodies, but it could do nothing with plasticity or with the host of other structure-sensitive properties on which the arts depend.

The geometry of perspective could be well handled by mathematicians, but the perspective of color could not be. The artist's intuitive knowledge of the psychology of perception has interacted strongly with science in the twentieth century, but for the most part science developed without art, and art was affected by science only through the changing world view that science promulgated, or indirectly through effects on technology.

The experience at moments of insight must be much the same among creative men of all kinds. However, the communication of new ideas, and especially their validation in terms that others will accept, is vastly different in different fields. As science and technology have become simultaneously broader in scope and more precise in individual purpose, their connection with art has become less and less apparent. Despite the austere and magnificent beauty of the order that is being uncovered by science, art has remained closer to technology than it has to science. As science has discovered the strength of simplicity, technology has become more complex. There is even a kind of aesthetic quality displayed by the interdependent relationships between the parts of an intricate machine, a complex process, or a large organization. Order per se is not art, and neither is complexity, but the finding of order in complexity is.

Looking back from the twentieth century, it is obvious that engineers, if not exactly aesthetes, have always had a rich and valid aesthetic experience in building their structures and devising their machines. A Newcomen engine at work with its massive rocking beam of oak mounted in a simple stone structure, with clanking chains and resonant iron bars, with its fiery furnace and jets of steam and its slow and irregular oscillation, *was* a work of art even if it was not consciously built as one or so appreciated in its own time. A modern artist, Garry Rieveschl, has recently proposed building a full-scale working Newcomen engine as a public monument, providing at once a reconstruction of a forgotten experience, a glance at a critical moment in technological history, and a reminder of the beauty and portentous quality of new contrivances.

In the theater, documentary drama is a similar art form based on a selective reconstruction of the past. Less conventional is the work of another Boston artist, Harris Barron, who has devised a performance which by effectively and unforgettably evoking the emotional experience of early aviators exemplifies the way in which art can extend human experience. Perhaps a poet on the first lunar landing would have done more for technology than an astronaut; certainly it will be poetic interpretations of space travel that will remain most in men's minds.

Symbolism in Art and Science

   Both art and science are basically symbol-making activities, and both have the quality of yielding metaphors that match far more than their creators intended. The scientist's equations and the conceptual models on which they are based often relate to other parts of nature which are mathematically similar but physically unrelated. This relationship is matched by that fundamental evocative quality of art, in which relationships developed by the artist with one aspect of form in mind turn out to suggest many other things to the eye of a viewer who has had different experiences. The artist consciously exploits the similarities in shape, color, texture, orientation, or other qualities of things of quite different natures; in fact, if there were not some such resonance, the viewer of a picture would find little to hold his attention. The scientist finds that a few basic patterns reappear at different levels and in different systems, but this is mainly because the types of interaction between the few units with which alone he can deal are, after all, quite limited: simplicity and symmetry do not allow many alternative arrangements.

   Historically, it is interesting to note occasions on which the decorative artist has developed designs that later were reinvented to represent important scientific concepts. One of the best examples of this is the use of circular mosaic tiles to build up two-dimensional polygonal patterns having all the characteristics of order, symmetry, and angular relationship between planes that are the basis of crystallography. This occurs in the Sumerian palace at Uruk, built in the middle of the fourth millennium B.C. (Fig. 24). Mosaicists ever since have been displaying examples of the combinatorial possibilities of simple geometric forms, none more magnificently than the Islamic tile workers of the fifteenth through the seventeenth centuries A.D.[65] The crystal lattice dislocation, which was conceived in 1926 and has become extremely important in solid-state physics, was modeled much earlier in the fitting of medieval suits of mail armor, in the studded decoration of Japanese cast-iron tea kettles, and with slight distortion in innumerable other repetitive designs.

   The best three-dimensional models of the close-packed-sphere arrangement of atoms on a crystal lattice occur in the famed granulation work of Etruscan goldsmiths in the sixth century B.C., though the technique was already 2,000 years old at that time. Figure 25 shows an octahedral ear ornament composed of tiny gold spheres, made in Persia in the ninth or tenth century B.C. Curiously, none of the Greek atomists hit on the basic principle that these things illustrate, namely, that the mere stacking of equal isotropic spheres would give rise to the directional anisotropy of crystalline matter; and it was left to Johannes Kepler in 1611 first to publish this principle in a scientific treatise—if that is the proper term for his playful essay inspired by the hexagonality of the snowflake.[66]

The stacked-ball model of the crystal is in every elementary textbook today; yet the idea proved difficult to accept and, despite Hooke's elaboration of the idea and Huygens's very effective use of it in explaining the properties of calcite crystals, it virtually disappeared for two and a half centuries for scientists preferred the concept of elementary polyhedra and, later, more elegant mathematical abstractions.[67]

Among the innumerable geometric patterns painted on early pottery in most cultures, there are many reminiscent of the magnetic-domain patterns of today's solid-state physicist. A more recent example of an artist's prescience lies in the work of the Dutch artist Maurits Escher, whose experiments with space-filling and repetitive patterns later provided the illustrations for an introductory book on symmetry for students of crystallography[68] and who in 1942 illustrated color-group symmetries "well before official crystallography even thought about them" and quite independently of their mathematical treatment by Schubnikov. Islamic tile workers had used them earlier, however, notably in the Alhambra (Fig. 28).

In recent years there have been many exhibitions and books relating scientific photographs to abstract art.[69] One of my own micrographs of a copper-silicon alloy once hung in the Museum of Modern Art! A particularly interesting coincidence is in some of the paintings of Piet Mondrian which were later found to have almost exactly matched the microstructures of some cubic crystals containing randomly nucleated plates of a precipitated phase growing at right angles to each other until interference.[70] Such correspondence, of course, is only possible in a period in which artists are unconcerned with representation of the human world and are, for whatever reason, seeking a simplicity commensurate with that of the physicist. Perhaps, however, sculpture and paintings with human symbolism will some day be found to have played a similar role in connection with psychological science. Op art certainly belongs in laboratories studying the simpler aspects of the physiology of perception.

Let us return to history. If the crystal lattice was slow of conception, the idea of the atom, of course, was not. Is there perhaps a connection between the use of pebble mosaics to depict human and animal figures in Greece in the fourth century B.C. and Leucippus's and Democritus's theories of matter? The concept that the distinguishing characteristics of matter arose in the shape, order, and orientation of parts in aggregation was certainly illustrated by the new mosaic forms, even if it was not suggested by them. At the present day, the printer's halftone is useful to illustrate information theory and discussions on structural hierarchy in matter.[71] And, again, in the same vein, is it absurd to suspect some connection between the revived receptivity to atomism at the end of the sixteenth century and the concurrent interest in the fine structure of a work of art that accompanied the new graphic methods? A rapid improvement in the quality of metal engraving accompanied the making of niello prints. Shading in both a woodcut and an engraving depends upon the control of discreet, nearly invisible lines which build up to a recognizable body. The painter, with continuous gradation of darkening and lightening even within a single brush stroke, does not need this kind of analysis; neither does the goldsmith with his repoussé bas-relief. Woodcuts in the nonatomistic Orient, exploiting mainly a variable quality of line and texture, are basically different from Western ones.

Today and Tomorrow

After this excursion into some of the past interaction between art and technology, it is tempting to speculate on their joint futures. It is fashionable today to note the similarity between the artist's creative insight and that of the scientist, but for some reason the technical side of art has been downgraded as "mere" technique. Yet the handling of matter will always be necessary to give reality to the artist's all-important vision. Without it he cannot influence the minds and feelings of other people. Moreover, since technique relates more closely to the everyday experience of most men and women, especially when they are young, an interest in it can provide a path to the deeper meaning of art and lead to an understanding of things that the intellectual has never been able to communicate. The artist, if not every art historian, has always known that technology is a basically important human activity.

The recent trend away from representational art in the Western world has, however, been accompanied by a perceptibly increased interest in techniques and materials and consequently by more widespread appreciation of the "minor" arts that make up such a major part of the archaeological record if not of art-historical writings.[72] This has inevitably been accompanied by an increased interest in Oriental art, for, while Western art tells mainly about individuals, ideas, and institutions, Oriental art (in accord with Oriental philosophy) tells more about nature—and it sensitively uses the properties of matter to do this. The subtle representations of the Chinese landscape painter arise from the properties of colloidal carbon and water interacting with the capillarity of various surfaces. The ceramist of the Far East can sometimes reveal the essence of things even better than can a painter because the ceramist's product constitutes in itself a direct example of the balance of natural principles, not merely a representation of them, selected and controlled by the potter just enough to invite appreciation by the human eye and hand.

The natural forces operating on matter are at last being consciously utilized in Western art: For example, Jēkabs Zvilna of Toronto has been producing two-dimensional patterns (visible only with optical enlargement) that result from the interaction of surface tension, shrinkage stresses, volatility, viscosity, and other physical forces working on selected substances under boundary conditions that he deliberately sets up. Though both the method and the results bear some relation to what ceramists have long done, especially in Japan, the nonnatural scale and the freer choice of materials gives his patterns a highly contemporary look. The recent "Kalliroscope" of Paul Matisse is a simple but perpetually fascinating device revealing shear gradients in a liquid containing floating micron-size reflecting platelets: patterns of endless variety and subtlety can be produced by controlling the conditions of turbulence or thermal convection, and close inspection even reveals to the naked eye the effects of molecular agitation (Brownian movement). Many other artists are experimenting with the aesthetic possibilities inherent in systems the details of which are fully determined by physical forces, but the boundary conditions are set by human intervention.

Many artists are currently exploring the properties of polarized light, of kinetic and balanced motion and flow, of simple magnetic interaction (Fig. 29), and of other phenomena which in the nineteenth century were used as rudimentary lecture demonstrations and laboratory experiments to evoke the interest of students in science.[73] It is high time that scientists admit that their experience in the laboratory is an aesthetic one, at times acutely so: the arid form of presenting their results has disguised this, and their respectable logical front often makes it invisible even to a student. The artist's interest in this aspect of science is very valuable. The introduction of scientific toys, under whatever name, to the general public and the opportunity to experience natural phenomena can only be applauded. The modern sculptor's skill in invoking viewer participation can aid enormously in the teaching of science, but his devices are usually of such simplicity that the initial feeling of pleasure cannot deepen by repetition into a rich aesthetic experience. The artists themselves rarely invite analysis of the intellectual overtones of their work and one sometimes suspects that they are unaware of them.

The visual excitement of the structures revealed by the microscope and electron microscope, of ion tracks in cloud chambers and interference patterns, has given rise to many fine exhibitions which have enriched the artist's vocabulary at the same time that they have heightened the scientist's sensibility. Yet it seems to me that in most of this the artist is just following others and is not fulfilling his particular role of revealing new significances in large, complex, perhaps social, patterns. Science is proliferating into more and more precise studies of more and more details. Higher energies beckon always away from the understanding of things on a human level to the smaller and simpler units of matter. So much knowledge has been acquired in this way that some scientists have claimed that no valid meaning can be established except by physical science. The most exciting frontier of biology has been on the molecular level, not life itself, which requires higher organization. After decades of neglect, however, something like old-fashioned natural history seems to be coming back into its own: the cell, biological form, and especially that comprehensive subject known as ecology, which is almost the art of science. Can the same thing happen in other areas? And can the artist, when he has learned some of the rudiments of science and technology, help?

Throughout history there has been a slow separation of art from the arts, and of science from both. As science became more definite, it became increasingly useful to technology, and it has given precision to both the design and the control of processes. With art, however, the very utility of its contributions to industrial design and advertising seems rather to have forced out the one component of it that is most needed, and today we are faced with the curious phenomenon of art being mainly a comment and a much-needed protest rather than a constructive suggestion of a way toward deeper understanding. Artists have found much to interest them in both the scientific and technological world, and they have shown that there is much beauty even in things such as galvanized iron roofing and the intricacies of stairs and piping in a chemical processing plant, to say nothing of the elegant patterns of electronic gadgetry. The strength of steel and concrete and the beauty of a streamlined surface are proper aesthetic experiences in today's world, and they become more so as artists explore their meaning. Many sculptors have learned to enjoy the properties of steel and to exploit the cutting and welding torch in producing sculpture. The role of the artist in pointing to common things and making one pause to look at them has always been important. He now plays a similar role in relation to science, not only in finding the visual delights of the New Landscape[74] but also in calling attention to experiences of the other senses that are possible in a scientific or technological environment.

Technology is by its very nature complex and thus is incapable of being completely understood. There are two kinds of simplification that can make this complexity handleable. The first is the scientists' recognition of the units and their interaction on a small precise scale and the other is the recognition of the connectivity of units—which sometimes is systems analysis but more constructively is art. As technology has passed from the individual work of craftsmen to an aggregate of integrated systems, the significance of individual processes has been lost precisely at the moment that they become most efficient. The discovery of new techniques owes less to artistic curiosity but now occurs in well-financed research laboratories and is increasingly dependent on science. Yet does not the transition from craftsman to technologist itself suggest a new area in which the artist should play a role? The new level of complexity in technology requires a new level of art, perhaps almost a social one. Indeed, the artist is needed now as never before, and only by an introduction of the artist's general sense of relationships will it be possible to restore the balance between social and individual needs. At least some of the artist's work will be devising schemes in which the pleasure of an intensely individual experience can interact with that of others to produce a more viable society than at present. The artist can highlight discrepancies and point up problems that should be solved before they become generally obvious, while it is the job of the technologist to say how to solve them, and to do so.

Here it should be noted that there are more possibilities of diversity above human scale than below it, and the dangers of oversimplification in social matters are correspondingly greater than in the realm of physics and chemistry. The more that individuals are able to enhance their differences without loss of contact, the richer their lives will be. Technology at last makes real diversity possible, but democratic egalitarianism is in danger of eliminating it. Part of the artist's job will be to oppose oversimplification in this world of immensely diverse possibilities. But needed beyond all else is an aesthetic feeling in the hearts and minds of technologists, who are so rapidly, at other peoples' behest, bespoiling the Old Landscape.

The author is grateful to his colleagues Arthur Steinberg and Heather Lechtman for many discussions in the general area of this paper, and to Julia Phelps, Janet Abramowicz, and Lynwood Bryant for critical comments on an earlier draft. His work in the history of technology is supported in part by grants from Mrs. Dominique de Menil, the Sloan Fund for Basic Research (M.I.T.), and the National Endowment for the Humanities. This paper has been preprinted in *Technology and Culture*, 11(1970), 491–549.

1. Leonardo da Vinci, *The Literary Works of Leonardo da Vinci . . .*, Ed. by Jean Paul Richter and Irma A. Richter (London, 1936), 1: 119.

2. *Ibid.*, 1: 25–26.

3. Robert Hooke, *Micrographia* (London, 1665), Preface (unpaginated).

4. Henri Focillon, *Vie des Formes* (Paris, 1947); English translation, *The Life of Forms in Art* (New York, 1948), *passim*, especially pp. 31–41, 76. The illustrations in the English edition are poorly selected to reinforce the author's points.

5. Arlette Leroi-Gourhan, *"Le Neanderthalien IV de Shanidar,"* *Bulletin de la Société Préhistorique Française* (*Comptes rendus séances mensuel*) (1968), *65:* 79–83. See also *New York Times*, June 13, 1968.

6. Though blue frit is characteristically an Egyptian product, the earliest examples of it are two frit vessels and some seals and amulets from Mesopotamia, in the Tall Halaf levels at Tall Arpachiyah near Ninevah, a period which lasted from roughly 4900 to 4300 B.C. (M. E. L. Mallowan and J. C. Rose, "Excavations at Tall Arpachiyah 1933," *Iraq* (1933–5), 2: 1–178. For a later appearance of it see Hans Wulff, et al., "Egyptian Faience: A Possible Survival in Iran," *Archaeology* (1968), 21: 98–107.

7. H. J. Plenderleith, "Metals and Metal Technique," in C. L. Wooley, *Ur Excavations, Vol. II, The Royal Cemetery* (London, 1934), 284–310, and Plates 138 and 162. For a discussion of early metallurgy, see T. A. Wertime, "Man's First Encounters With Metallurgy," *Science* (1964), 146: 1257–67, and C. S. Smith, "Materials in Civilization and Science," *Science* (1965), 148: 908–917. The best comprehensive history of metallurgy is that by Leslie Aitchison, *A History of Metals* (2 vols., London, 1961).

8. Theophilus, *De Diversis Artibus*, manuscript treatise, *ca.* 1123 A.D. Latin text and translation by C. R. Dodwell (London, 1961); translation with technical notes by J. G. Hawthorne and C. S. Smith (Chicago, 1963). Chapters 85–87 deal with bell casting.

9. Vannoccio Biringuccio, *De la Pirotechnia*, (Venice, 1540). English translation by C. S. Smith and M. T. Gnudi (New York, 1942), 255–60.

10. Noel Barnard, *Bronze Casting and Bronze Alloys in Ancient China* (Canberra and Nagoya, 1961); R. J. Gettens, *The Freer Chinese Bronzes*, Vol. II of *Technical Studies* (Washington, D.C., 1970).

11. S. Delbourgo, *"L'étude au laboratoire d'une statue découverte à Agde,"* *Bulletin du Laboratoire du Musée du Louvre* (1966), 7–12; H. Lechtman and A. Steinberg "Bronze Joining—A Study in Ancient Technology," in *Art and Technology: A Symposium on Classical Bronzes*, ed. by S. Doeringer et al (Cambridge, Mass., 1970), 5–35.

12. Hans Drescher, *Der Überfangguss* (Mainz, 1958).

13. R. J. Gettens, "Joining Methods in . . . Ancient Chinese Bronze Ceremonial Vessels," in *Application of Science in Examination of Works of Art*, ed. by W. J. Young (Boston, 1967), 205–17.

14. The history of the blowpipe has yet to be written. Blowing through pipes to urge charcoal (?) fires for smelting and melting doubtless preceded the use of bellows. The small mouth blowpipe with a lamp or candle was used by jewelers for local soldering operations and in the seventeenth century was suggested for testing ores. Comprehensive schemes of chemical analysis based on it were developed in Sweden in the last half of the eighteenth century, but were slowly displaced, except in the field, by wet methods of analysis. Blowpipe analysis was regarded as an essential part of the training of a young chemist until very recently, and the author still traces his feel for the nature of very many chemical substances and reactions to his work with the blowpipe as a schoolboy.

15. Hans Wulff, *et al.* (n. 6, above).

16. C. S. Smith, "Porcelain and Plutonism," in *Toward a History of Geology: Proceedings of the New Hampshire Conference on the History of Geology, 1967*, ed. by C. J. Schneer (Cambridge, Mass., 1969), 317–38.

17. Cennino d'A. Cennini, "On the Character of Ultramarine Blue and How to Make It," *Il Libro del'Arte* [*ca.* 1400, A.D.], ed. and trans. by D. V. Thompson, Jr. (New Haven, 1933), Chapter 62. A more complete account of the flotation process was recorded slightly later in the Bologna manuscript reported and translated by Mary P. Merrifield, *Original Treatises Dating from the XIIth to XVIIIth Centuries on the Arts of Painting in Oil* (2 vols., London, 1849).

18. C. S. Smith, "The Early Development of Powder Metallurgy," *Powder Metallurgy*, ed. by John Wulff (Cleveland, 1942), 4–17; Shirley Alexander, "Medieval Recipes Describing the Use of Metals in Manuscripts," *Marsyas* (1966), *12*: 34–51. Shirley Alexander, "Base and Noble Metals in Illumination," *Natural History* (1965), *74* no. 10:31–39.

19. S. Giedion, *Space, Time and Architecture* (Cambridge, Mass., 1953); Norman Davey, *History of Building Materials* (London, 1961); L. F. Salzman, *Building in England down to 1540* (Oxford, 1952); Marion E. Blake, *Ancient Roman Construction in Italy* (Washington, D.C., 1947).

20. A. G. Drachmann, *The Mechanical Technology of Greek and Roman Antiquity* (Copenhagen, 1963); B. S. Brumbaugh, *Ancient Greek Gadgets and Machines* (New York, 1966).

21. Alfred Chapuis and Edmond Droz, *Automata: A Historical Account and Technical Study* (Neuchatel, 1958). See also *Les automates dans les*

Oeuvres d'Imagination (Neuchatel, 1947) and several other works by Chapuis.

22. V. Gordon Childe, "Rotary Motion," in *A History of Technology*, ed. by Charles Singer et al. (London, 1954), 1:187–215.

23. The apogee of ornamental turning and its gadgetry is recorded in John Jacob Holtzapffel, *Turning and Mechanical Manipulation*, vol. 5 of *The Principles and Practice of Ornamental or Complex Turning* (London, 1884). The book by A. K. Snowman, *Eighteenth Century Gold Boxes of Europe* (London, 1966), illustrates innumerable surfaces, both enameled and plain, whose decorative charm derives directly from engine turning.

24. Robert S. Woodbury, *A History of the Lathe to 1850* (Cleveland, 1961); L. T. C. Rolt, *A Short History of Machine Tools* (London and Cambridge, Mass., 1965).

25. James Mellaart, *Çatal Huyuk; A Neolithic Town in Anatolia* (London and New York, 1967).

26. T. F. Carter and L. C. Goodrich, *The Invention of Printing in China and its Spread Westward*, 2nd ed. (New York, 1955).

27. The earliest description of typecasting is in V. Biringuccio *De la pirotechnia* (Venice, 1540), where it appears appropriately in the chapter on the pewterer's art.

28. Hellmut Lehmann-Haupt, *Gutenberg and the Master of the Playing Cards* (New Haven and London, 1966).

29. A. M. Hind, *Nielli, Chiefly Italian of the XV Century: Plates, Sulphur Casts and Prints Preserved in the British Museum* (London, 1936). For reproductions of other sulphurs and niello prints, see John G. Phillips, *Early Florentine Engravers and Designers* (Cambridge, Mass., 1955).

30. James G. Mann, "The Etched Decoration of Armour," *Proceedings of the British Academy* (1942), 28: 17–44.

31. Sidney Goldstein, in a private communication, reports that fragments of hammered gold that had unmistakably been cemented were uncovered in a 550 B.C. workshop site at Sardis. They were found during the 1968 campaign of the Fogg Museum at Sardis. The earlier Lydian coins were of unrefined electrum.

32. Emil W. Haury, "Etched shells," in *Excavations at Snaketown*, ed. by H. Gladwin et al., *Medallion Papers*, vol. 25 (Tucson, 1937), 148–53.

33. H. Lechtman, "The Gilding of Metals in Precolumbian Peru," in *Application of Science in the Examination of Works of Art* (Seminar III, June, 1970) ed. by W. J. Young (Boston Museum of Fine Arts, in press). The gilding process is actually the selective superficial corrosion of silver and copper from a relatively dilute alloy of gold, using an acid ferric sulphate mineral. The presence of reserved areas of silver in these objects has not previously been noted, although there have been speculations on welded bimetallic construction.

34. H. R. E. Davidson, *The Sword in Anglo-Saxon England* (Oxford, 1962). On the metallurgy of the pattern-welded blades, see C. S. Smith, "The Pattern Welded Blade," in *A History of Metallography* (Chicago, 1960), Chapter 1, and the references cited therein.

35. C. S. Smith, "The Damascus Blade," in *A History of Metallography* (Chicago, 1960), Chapter 3; C. Panseri, "Damascus Steel in Legend and in Reality," *Gladius* (1965), 4: 5–66.

36. B. W. Robinson, *The Arts of the Japanese Sword* (London, 1967); C. S. Smith, "A Metallographic Examination of Some Japanese Sword Blades," *Doc e Contributi per la Storia della Metallurgia* (Milan, 1957), No. 2: 42–68.

37. *Compositiones ad tingenda musiva* . . . , ed. and trans. by H. Hedfors (Uppsala, 1932); Thomas Phillipps, ". . . A Manuscript treatise . . . entitled Mappae Clavicula," *Archaeologia* (1847), 32: 183–244; Wilhelm Ganzenmuller, "Ein unbekanntes Bruchstück der Mappae Clavicula aus dem Anfang des 9. Jahrhunderts," *Mitteilungen zur Geschichte der Medezin der Naturwissenschaft der und der Technik* (1941), 40: 1–15; a translation of the *Mappae clavicula* by J. G. Hawthorne and C. S. Smith is to be published.

38. Conrad Kyeser, *Bellifortis*, facsimile, transcript and German translation by G. Quarg (2 vols., Dusseldorf, 1967). Versions of similar recipes without, however, using distillation to make strong acid, appear in several 15th century sources. The first printed account of etching is in the anonymous Dutch *T. Bouch vā Wondre* (Brussels, 1513, reprinted with commentary by H. G. T. Frencken, Roermund, 1934). Next is the important little pamphlet *Von Stahel und Eysen* (Nuremberg, 1532) which was reprinted many times both by itself and with other material in the series of *Kunstbüchlein* and other books of secrets. For a modern English translation, see C. S. Smith, *Sources for the History of the Science of Steel 1532–1786* (Cambridge, Mass., 1968), 37–38.

39. L. Reti, "How old is hydrochloric Acid?" *Chymia* (1965), 10: 11–23; Sidney Edelstein and H. C. Borghetty, eds., *The Plictho of Gionventura Rosetti* (Cambridge, Mass., 1969); G. B. della Porta, *Magiae naturalis libri viginti* (Naples, 1589). The 1658 anonymous English translation of the section "How to Grave Porphyr Marble without an Iron tool" is reproduced in C. S. Smith, *Sources for the History of the Science of Steel 1532–1786* (Cambridge, Mass., 1968), 37–38. See also Haury (n. 32, above).

40. C. S. Smith, "The Discovery of Carbon in Steel," *Technology and Culture* (1964), 5: 149–75.

41. C. S. Smith, "*Moiré métallique*," in *A History of Metallography* (Chicago, 1960), 63–65. A later description of the process in an American hardware catalogue is quoted by H. J. Kaufman, *Early American Ironware, Cast and Wrought* (Rutland, Vt., 1966).

42. Alexis [pseudonym], *Secreti . . . del Alessio Piemontese* (Venice, 1555). There were innumerable subsequent editions and translations of this book, which is the most complete of all the early books of recipes for artists, craftsmen and housewives.

43. For a history of nature printing, see Ernst Fischer, "Zweihundert Jahre Naturselbstdruck," *Gutenberg Jahrbuch* (1933), 186–213.

44. Carl von Schreibers, *Beyträge zur Geschichte und Kenntniss meteorischer Stein- und Metall-massen* (Vienna, 1820). See also C. S. Smith, *A History of Metallography*, 150–56; and "Note on the History of the Widmanstätten Structure," *Geochimica et Cosmochimica Acta* (1962), 26:271–72.

45. Alois von Auer, *Der polygraphische Apparat* (Vienna, 1853). This includes a portfolio of fine prints made by all methods of reproducing illustrations then known: several are of scientific subjects.

46. Norman Higham, *A Very Scientific Gentleman: the Major Achieve-*

ments of Henry Clifton Sorby (Oxford, 1963). For Sorby's work on steel, see C. S. Smith, *A History of Metallography*, 169–85; A. R. Entwisle, "An Account of Exhibits Relating to Henry Clifton Sorby," *Metallography 1963* (London, 1963); and papers by C. S. Smith, D. W. Humphries, and Norman Higham in *The Sorby Centennial Symposium on the History of Metallurgy* (New York, 1965).

47. Gustav Alexander, *Herrengrunder Kupfergefässe* (Vienna, 1927).

48. Thomas Hale, *An Account of Several New Inventions or Improvements now Necessary for England* . . . . (London, 1691).

49. The early books on electroplating devote much space to the advocacy of one or other view of its history. See particularly Alfred Smee, *Elements of Electrometallurgy* (London, 1842, 2nd edition 1843); George Gore, *The Art of Electrometallurgy* (London, 1877). The process of electrotyping, commercially used in 1840, spread rapidly across the Atlantic (see figure 15).

50. Alfred Bonnin, *Tutenag and Paktong* (Oxford, 1924).

51. R. A. F. de Réaumur, *L'art de convertir le fer forgé en acier* . . . . (Paris, 1722). English translation by A. G. Sisco (Chicago, 1956), 340–59.

52. R. A. F. de Réaumur, "L'art de faire une nouvelle sorte de porcelaine, . . . ou de transformer le verre en porcelaine," *Mém. Acad. Sci.*, 1739 (published 1741), 370–88. Réaumur's porcelain is extremely rare, but a box that seems to be made of it has recently been described by R. Strasser, *Journal of Glass Studies*, 1967, 9: 118. See also n. 16, above.

53. The lead rolling mill is illustrated in Jost Amman's woodcut of the glazier (Jost Amman, illustrator, *Eigentliche Beschreibung aller Stände auff Erden . . . durch . . . Hans Sachs* [Frankfurt, 1568]). The same blocks were used in a Latin edition in the same year. Plates showing the profiled iron shapes and some new window designs that they made feasible appear in [–. Bullot] *Mémoire sur les ouvrages en fer et en acier qui se fabriquent dans la manufacture Royale d'Essonne par le moyen du laminage* (Paris, 1753). For more detail see C. S. Smith, Architectural shapes of hot-rolled iron, 1753," *Technology and Culture* (1971) 12: in press.

54. C. S. Smith, "Matter versus Materials: A Historical View," *Science* (1968), 162: 637–44.

55. A. J. Hopkins, *Alchemy: Child of Greek Philosophy* (New York, 1934).

56. Lynn White, Jr., "The Iconography of Temperantia and the Virtuousness of Technology," in *Action and Conviction in Early Modern Europe*, ed. by T. K. Rabb and J. E. Seigel (Princeton, 1969), 197–219.

57. Three useful collections of paintings and other works illustrating technological scenes and devices are Heinrich Winkelmann, *Der Bergbau in der Kunst* (Essen, 1958); Vaclav Husa et al., *Traditional Crafts and Skills* (Prague & London, 1967); F. D. Klingender, *Art and the Industrial Revolution* (London, 1947). Many works on the history of technology reproduce artists' works as illustrative material. Emil E. Ploss, *Ein Buch von alten Farben* (Heidelberg and Berlin, 1962), as befits its subject, is an unusually fine mixture of historical and artistic material.

58. Helmuth T. Bossert and Willy F. Storck, eds., *Das mittelalterliche Hausbuch.* (Leipzig, 1912).

59. Conrad Kyeser . . . , *Bellifortis*. Facsimile, edited and translated into German by Götz Quarg (2 vols., Dusseldorf, 1967).

60. Amman (n. 53, above).

61. John Read, *Prelude to Chemistry* (London, 1936); Read, *Humour and Humanism in Chemistry* (London, 1947).

62. Giorgio de Santillana, "The Role of Art in the Scientific Renaissance," in *Critical Problems in the History of Science*, ed. by Marshall Clagett (Madison, 1959). Reprinted with other essays in *Reflections on Men and Ideas* (Cambridge, Mass., 1968). See also the essays in H. H. Rhys (ed.), *Seventeenth Century Science and Arts* (Princeton, 1961).

63. G. Evelyn Hutchinson "Psychological and Aesthetic Factors in the Progress toward Realism, A.D. 1280–1480" (Paper presented at Symposium on Art and Science, AAAS meeting, December, 1968).

64. Stillman Drake, "Renaissance Music and Experimental Science," *Journal of the History of Ideas*, (1970), 31:483–500.

65. Edith Muller, *Gruppentheoretische und strukturanalytische Untersuchungen der maurischen Ornamente aus der Alhambra in Granada* (Rüschlikon, 1944).

66. Johannes Kepler, *Strena seu de Nive Sexangula* (Frankfurt, 1611); translated by Colin Hardie (Oxford, 1966). Though the symmetry (not always sixfold!) of the snowflake appears commonly enough on today's Christmas cards, its decorative qualities do not seem to appear in art until after its depiction in scientific works. In the Far East, window lattice patterns representing interfering ice crystals on a frozen pond are common. [Daniel S. Dye, *A Grammar of Chinese Lattices*, (2 vols., Cambridge, Mass., 1937)], but the earliest oriental use of true snowflake symmetry appears to be that on a Japanese sword guard made by Harukiro Hirata in 1828, obviously related to the drawings (Fig. 26) that were being made by Toshitsura Doi under the influence of Dutch science and published five years later. [Toshitsura Doi, *Sekka zusetsu* (Illustrations of snow crystals) (Tokyo, 1833; suppl., Tokyo, 1840; reproduced with commentary and English summary by Teisaku Kobayashi, Tokyo, 1968)]. Figure 27 shows an elegant iron guard made somewhat later by a member of the famed Goto family.

Doi's drawings have a symmetry that is quite un-Japanese. Virtually every drawing of a snowflake that has been published, whether in a work of art or of science, depicts almost exact symmetry, reflecting the unwarranted but firm belief in the precision of nature and the inability of the eye to see the unexpected. A glance at any photograph, or, better, the flakes themselves, will show many small differences between the six dendiritic branches of even the best flake. And, of course, most snow falls as irregular aggregates displaying no symmetry whatever.

67. John G. Burke, *Origins of the Science of Crystals* (Berkeley, 1966).

68. Caroline H. Macgillavry, *Symmetry Aspects of M. C. Escher's Periodic Drawings* (Utrecht, 1965); M. C. Escher, *Grafiek en tekengen* (Zwolle, 1960); English translation, *The Graphic Work of M. C. Escher* (London, 1961, 2nd ed., enlarged, 1967).

69. See, for example, Lancelot L. Whyte, ed., *Aspects of Form: A Symposium on Form in Nature and Art* (London, 1951); *The Artist Looks at the Scientist's World*, catalogue of exhibition organized by the Renaissance Society, University of Chicago, 1952; Gyorgy Kepes, ed., *Structure in Art and Science* (London and New York, 1965); Philip C. Ritterbush, *The Art of Organic Form* (Washington, D. C., 1968); Georg Schmidt, ed., *Kunst und*

Naturform: Form in Art and Nature (Basel, 1960); Paul Weiss, "Beauty and the Beast: Life and the Rule of Order," *Scientific Monthly* (1955) 81: 286–99.

70. R. W. Cahn, "Art in Science, Science in Art," *Museum* (UNESCO) (1968), 21: 16–21.

71. Paul A. Weiss, "1 + 1 ≠ 2," in *Neurosciences: A Study Program*, ed. by Gardner C. Quarton, Theodore Melnechuk, and Francis O. Schmidt, (New York, 1967), 801–21.

72. It is interesting to note that the published catalogues of three recent art exhibitions have long technical introductions: Herbert Hoffmann and Patricia Davidson, *Greek Gold* (Boston, 1965); Dominique de Menil, *Made of Iron* (Houston, Texas, 1966); David G. Mitten and Suzannah H. Doeringer, *Master Bronzes of the Classical World* (Cambridge, Mass., 1967).

73. Jack Burnam, *Beyond Modern Sculpture* (London and New York, 1968).

74. Gyorgy Kepes, *The New Landscape in Art and Science* (Chicago, 1956); *idem*, (ed.), *Structure in Art and Science* (New York, 1965).

Commentary on the Paper of Cyril S. Smith

By Melvin Kranzberg, Case Western Reserve University

CYRIL STANLEY SMITH is rightly considered one of our major historians of science and technology. This excellent paper on the interactions among art, science, and technology strikes me as one of the most significant of his scholarly researches, for it enlarges—almost in a quantum leap—the dimensions of our discipline. Dr. Smith's thesis undoubtedly derives from his concern with structure (itself an outgrowth of his previous and concomitant career as a distinguished metallurgist), his feeling for texture, and his metallographic studies of ancient art objects. And despite its profound significance for the history of science and technology, Dr. Smith's is a simple thesis, namely, the importance of the aesthetic motivation—the joy of aesthetic satisfaction—in the development of science and technology.

I submit that this is a most revolutionary thesis. It is revolutionary, not in the sense that it overturns previously existing explanations for the development of science and technology, nor because it rules out all preceding explanations. Nor is it revolutionary because no one has ever hinted at it before. Other scholars—Lewis Mumford, John Kouwenhoven, Carl W. Condit, to name but a few—have suggested the importance of the aesthetic element in leading man to investigate nature and improve techniques, and in shaping the outlines of specific scientific theories and technological artifacts, but they have not made it their central argument. Dr. Smith, however, has now advanced the aesthetic hypothesis as the thesis of a major paper; he has stated his thesis cogently, succinctly, and what is more important, he has provided some irrefutable documentation.

This paper is remarkable, then, in the vigor, clarity, and proof with which Dr. Smith has advanced his thesis. It is revolutionary in that it enlarges the dimensions of the study of the history of science and technology by adding a new parameter to our concerns. No historian of science or technology who is worth his salt can now discuss developments in either of those fields without some consideration of the aesthetic stimulus to scientific and technological advance. This I call a major contribution.

Dr. Smith's emphasis on the close relations between art and the technology of materials brings particular satisfaction to historians of technology. For too long we have been envious of our brethren in the history of science. They perform their research in libraries with the rich smell of vellum and fine leather bindings; they keep the intellectual company of Oxford dons, Cambridge divines, and intimates of popes and princes; and they deal with such rarefied matters as the macrocosm and the microcosm. Historians of technology, on the other hand, must figuratively inhabit grimy mine pits, stinky chemical works, and noisy factories; our companions for our intellectual excursions are, for most of history, lowly-born slaves, horny-handed peasants, and greasy, sweaty mechanics; and we concern ourselves with inanimate machines, inarticulate or secretive craftsmen, and the mundane considerations of daily working life.

Now Dr. Smith elevates the position of our intellectual concerns; he tells us that the development of technology, even more than that of the scientists, is tied directly to high culture, to the fine arts. Small matter to us that our ties are chiefly with the decorative arts; we are grateful for this new-found status, this humanizing of the motivations of technological advance to include not only the grubby pursuit of material gain but also to comprehend high aesthetic purpose, even if only in the so-called "minor" arts.

Assuming at least the partial validity of Dr. Smith's thesis, I find that it has two additional significant implications for the history of science and technology. The first of these is by no means new, but it needs constant reiteration. It is the importance of socio-cultural elements in the development of science and technology. Dr. Smith tells us that the discovery of many phenomena utilized by modern metallurgists is very old. For example, an examination of the jewelry and other metal objects from the famous royal graves at Ur (ca. 2600 B.C.) reveals that their makers possessed knowledge of virtually every type of metallurgical phenomenon except the hardening of steel that would be exploited by technologists in the entire period up to the end of the nineteenth century A.D.

The question immediately occurs: why for so many centuries was there not industrial exploitation of the knowledge of metals apparent in these Ur tombs of the third millennium B.C.? Failure to exploit this knowledge fully can only be explained in terms of the social and economic and cultural factors at work over the span of some 4,500 years.

While the importance of these factors has long been recognized, Dr. Smith has really added a new element. Heretofore these socio-cultural factors were used to explain, at least partially, why and how certain scientific and technological questions came to be investigated at certain times, and why and how the answers to these investigations took the form which they did. But Dr. Smith points out that, in the case of materials at least, the knowledge already existed at the dawn of the historical era. The historical question then becomes one of examining the socio-cultural factors to explain why this knowledge was not exploited more fully to affect the lives of people throughout these forty-five centuries.

Dr. Smith himself implies a possible answer to these questions, and this, I think, represents another important implication. To put it quite simply, I think that he is here laying the foundations for what might be termed an aristocratic theory of the development of technology. Let me explain.

He asserts that the royal burial objects at Ur are far from being representative samples of contemporary use of any material; the court would appropriate the best work to its own ends. He then states that the invention of a technique is more likely to occur in an aesthetic environment than in a practical one, and he claims that in antiquity the failure of advances in mechanical devices to parallel those in materials lies in the fact that the same aesthetic rewards were not present in the mechanical field. And when mechanical devices finally did develop at a later date, particularly machine tools, their earliest use, too, was for aesthetic objects.

Dr. Smith has shown the connection between technical innovations and the aristocratic employment and enjoyment of art. Since the technology of doing work—the lower-class technology—had few aesthetic elements which appealed to the tastes of the aristocrats, innovation in such mundane fields would not be encouraged in a society dominated by an aristocracy or oligarchy. Although Dr. Smith implicitly denies Benjamin Farrington's thesis that slave labor was responsible for the failure of the Greeks and Romans to appreciate the advantages of mechanical power, at the same time he implicitly confirms the class structure as a factor in technological development by imputing the failure to develop mechanical devices in the ancient world to the lack of aesthetic rewards. Unwittingly perhaps, Dr. Smith offers us a new theory of technological stimulation based on an aristocratic desire for decorative and expensive objects.

This situation apparently prevails until the eighteenth century. The aristocratic orientation of pre-eighteenth-century technological development can be seen in Dr. Smith's assertion that Réaumur specifically states that his incentive was to provide cheaply for the masses decorative objects of a kind expensively made and previously available only for the rich. Thus the mid-eighteenth century ushers in an era of mass production for the masses—the Industrial Revolution—a more democratic technology.

Perhaps I am inferring too much from Dr. Smith's argument. I do not mean that the lower classes had no aesthetic appreciation, but certainly their tastes must have been coarser than those of the aristocracy. The materials techniques described by Dr. Smith were for the production of a small number of beautiful objects which could only be afforded by the upper levels of society, the nobles, the luxury trade, the church—in brief, a small closed "establishment"—and I believe that this represents an aristocratic theory of the development of technology.

Dr. Smith's paper suggests that we are perhaps on the verge of a new theory—a class basis for interpreting the direction of technology which is independent of the Marxist dialectic. While Lilley might have been correct in calling iron "the democratic metal," Smith's theory helps explain why metallurgical innovations had an aristocratic bias until modern times, and why almost three thousand years elapsed between the introduction of the so-called "democratic" metal and the actual coming of democracy.

There is, of course, a possible technological explanation for the failure to exploit these techniques for industrial purpose. That explanation would be based upon the differences in scale between limited production for luxury use and mass production for public consumption. In other words, while the jewelers and craftsmen could make innovations in decorative articles for the rich, they lacked the knowledge and means to scale-up these techniques for industrial applications. I suspect that it was not that, however, but rather the lack of motivation in an aristocratically-dominated society. Motivated by aesthetic stimuli and producing for a small-scale luxury market, the whole socio-cultural milieu of an aristocratic or oligarchic society militated against the wider exploitation of these techniques until a more democratic society began coming into existence in the mid-eighteenth century.

I am by no means certain that the aristocratic theory of technological innovation which I have imputed to Dr. Smith, or at least inferred from his paper, can be sustained. But I claim that it is a fertile hypothesis which might be extremely useful in explaining the nature, character, and direction of much technological innovation for a period of some 5,000 years, until relatively recent times. It is certainly a thesis which suggests further investigation.

Historians, too, have an aesthetic sense, a feeling for proportion, form, balance, and symmetry. Having thus paid tribute to the major contribution made by Dr. Smith, I feel that I must now balance these favorable judgments by references to certain portions of his paper which I think are somewhat faulty. However, my criticisms deal with matters which are only of secondary concern in Dr. Smith's treatise; they by no means affect the major points which he has made so effectively.

One fairly serious omission is his lack of any sustained argument on how technology has influenced art in addition to the techniques, tools, and materials used by the artists. With the exception of the Newcomen engine, Dr. Smith tells us little about how works of technology might themselves be considered objects of art, and how technology has influenced art in furnishing subject matter, themes, and approaches. Furthermore, technology has provided media for communicating art, and it has helped to create a society which the artist can either embrace or rebel against. Dr. Smith alludes to all these matters, but I wish that he had given them more systematic treatment.

I find his section on "Symbolism in Art and Science" the least satisfactory part of his paper. Dr. Smith correctly points out that both art and science are "symbol-making activities." He then goes on to point out that the decorative artist has developed designs that later were re-invented to represent important scientific concepts. I doubt if this shows a meaningful relationship. Perhaps the fact that they both utilize the same symbols does not indicate connections between art and science so much as it does the tendency of the human mind to symbolize in the same terms. The fact that certain crystalline structures resemble the geometric patterns of artists does not prove that the interrelationships of art and the sciences are anything but coincidental. If there is something meaningful in this coincidence, what is it? Does art imitate nature, or vice versa?

It might have been more meaningful if Dr. Smith had examined another facet of the relationships among science, technology, and art. Can it not be said that all three—science, art, and technology—share the human propensity for making order from chaos?

There are other lines of connection among science, technology, and art—psychological, philosophical, and aesthetic—which Dr.

Smith might have investigated further. But he has contributed so much in what he has written that we cannot fault him for not dealing in depth with every possible interrelationship among these three vital human activities.

Finally, Dr. Smith closes his paper with a series of observations regarding contemporary and future relationships. In a sense, he concludes with a plea that C. P. Snow's "two cultures" be brought together, with the artist helping to meet some of the problems of our times by learning some of the rudiments of science and technology, and with the technologists acquiring an aesthetic feeling in their hearts and minds.

I must confess that I have little hope that artists will study the calculus, comprehend quantum theory, or even learn the fundamentals of electronic devices. Much more promising is the possibility of developing the aesthetic sensibilities of the technologists, especially since Dr. Smith has provided us with examples ranging throughout history of the aesthetic motivations of technologists. Besides, nowhere could we find a better symbol and representative of this welding together of aesthetics, science, and technology than a Cyril Stanley Smith himself, as he has so amply demonstrated in this presentation.

Commentary on the Paper of Cyril Stanley Smith

By Carl W. Condit, Northwestern University

I THINK ONLY Cyril Smith has the learning, the wit, and the insight to write a paper on the whole sweep of the historical interactions of art, science, and technology. Only he could present such an immense panorama with the necessary erudition but at the same time with a light touch and a lively spirit, free of pedantry, full of contagious enthusiasm arising from his admiration for the artist and craftsman and from his delight in their work.

The essential theme of his essay is summed up in a single sentence: "The scientifically important properties of matter and the technologically important ways of making and using them have been discovered or developed in an environment which suggests the dominance of aesthetic motivation."

I have no serious quarrel with this thesis, and I am willing and prepared to extend it with further illustrations. It implies that art is primary, that technology to a considerable degree evolves in the wake of artistic demands or is intimately bound up with them, and that science only after a very long period of development arrives at a stage where it can say something illuminating about the technically useful properties of matter.

Actually, scientific analysis will always fall short of explaining the multi-layered truths of nature communicated through the experience of the artist-craftsman. Cyril Smith suggests this in another way when he says, "Since technique relates more closely to the everyday experience of most men, it can provide a path to the understanding of things the intellectual has never been able to communicate."

If I have any criticism of this thesis, it is only that I find it somewhat oversimplified, with missing elements at both ends of the historical spectrum, the primitive phase of pure aesthetic involvement in natural materials, and the ultimate phase in which craftsmanship has given way to predictive mathematical science.

There is no question that on the highest levels the combined powers of artistic creativity and technological skills have until very recently transcended the powers of scientific analysis to elucidate the inner character of natural processes and the human spirit. For example, only in the past three years has it been possible to determine the internal stress pattern in the structural system of a Gothic cathedral, and then only with the most advanced techniques of photoelastic model analysis (experiments at Princeton University by Robert Mark and colleagues in connection with David Billington's course in study of past works of structural art for graduate students in civil engineering). But model analysis, however expert, entails an extreme reduction in scale and the substitution of a transparent plastic for masonry, with the result that the stressed model bears the same relation to the original as a map does to the actual surface of the earth. A very good analogy, it seems to me, to the relation between the symbolism of science and the actual processes of nature. Through the Princeton experiments we gained insight into the techniques of the past and understood the validity of the structural forms involved; but because of the nature of the analytical method, the multidimensional synthesis is revealed only in the work itself.

But I am getting ahead of my own theme, so let's go back to the original point. I would like to reformulate Cyril Smith's thesis, then try to explore the interaction of art, techniques, and science by analyzing a particular work drawn from the building arts of the ancient world.

The assertion that the technical exploitation of material properties arose in an environment marked by the dominance of aesthetic motivation is true only because of the dominance of a prior motivation—namely, the need to understand, explain, and interact with the world in terms comprehensible and congenial to the human spirit. Art has always been a representation, in whole or in part, of what we might call, depending on the age and the milieu, the mythopoeic or apocalyptic or phenomenal or immediately intuitive view of the world, and it is essential to the rituals by which man enters into the cosmic drama of which he is a part.

Thus art, in one way or another, is made up of symbolic images of a cosmos (in the literal Greek sense), an encompassing or an internal order, first described in myth, later in theology and metaphysics, ultimately in psychology and literature. So we arrive at a kind of sequence, partly historical, partly transcending history: myth—art—the technical embodiment—science. But the physical embodiment itself adds something that the artistic idea alone cannot give.

Let me present this in concrete terms with respect to a specific example, one of the greatest works of building art in the ancient world, namely, the Pantheon in Rome, built in 118/119–125/128, in the reign of Hadrian, the Renaissance prince or Thomas Jefferson of the Roman emperors—soldier, administrator, poet, painter, architect, town planner. The example offers certain advantages because it appears in a relatively simple scientific age but is a highly sophisticated work of building art.

The Pantheon is one of the earliest of the great Roman buildings that belong to the category of form-resistant structures. These begin to appear in the revolutionary period of the late first century–early second century, during the reigns of Nero, Trajan, and Hadrian. They were made possible by the understanding of the physical properties and the structural potentialities of a revolutionary material —hydraulic concrete. The cementing agent, lime, had been long known; the hydraulic agent, called *harena fossicia* by Vitruvius and *pulvis Puteolorum* by Pliny, a sandy volcanic earth made up chiefly of the oxides of iron, aluminum, and silicon, was a Roman discovery.

All the recorded structural theory and constructional techniques known at the time are in Vitruvius, but these fall far short of the theory that is embodied in the Pantheon. As a matter of fact, full theoretical understanding of the internal structural action was not really known until the twentieth century. (Vitruvius is exasperating because of what he omits and because of his incomprehensible explanations.)

The Pantheon was designed to meet the functional need of maximum open interior space and to provide a symbolic image of a specific religious and metaphysical world view. That the second took precedence over the first is unmistakably revealed by a number of characteristics:

(1) The structure consists of a hemispherical dome supported by a cylindrical drum with a height equal to the radius of the sphere. (2) If the sphere were completed its lower half would be exactly contained within the circumscribed cylinder. (3) Thus the internal diameter of dome and cylinder, 142 English ft. (147 Roman ft., suggesting a basic module of 150 Roman ft.) equals the height from the floor to the crown. (4) Further, the complex of cylinder and hemisphere would fit exactly into a cube. (5) The location and diameter of the oculus suggest that the Pantheon is based on the concept of an analemma (Vit., IX, vii, viii), a geometric device designed to show the sun's declination throughout the year.

If I understand this perplexing and controversial matter, the oculus allows the sun's rays to reach the floor at the meridian passage only within a small period around the summer solstice. But the diameter of the oculus seems to have been determined by the latitude of Syracuse, suggesting an antecedent in that city. Puzzling.

The conclusion is clear: the Pantheon is symbolic and ritualistic, an anagogical image of the spherical Pythagorean-Platonic cosmos. The physical embodiment of this image on the scale demanded by Hadrian required a building with a clear span far exceeding anything in the past and not to be exceeded in the future until the mid-nineteenth century, and one, furthermore, that would be economically and structurally feasible.

The natural material was concrete, brick-faced for aesthetic purposes, and used for the cylinder, dome, and intermediate block, only the entrance portico being stone masonry. The technical solution shows a profound qualitative understanding of the structural action of a dome and the associated stress distribution, although this theory does not appear in a mathematico-scientific form until the late nineteenth century.

Chief structural elements, from floor to crown are: (1) Continuous foundation ring of concrete with travertine aggregate. (2) Drum consisting of eight radial piers rendered into a smooth external cylinder by relatively thin walls across the voids between them. (3) Brick conical vaulting, with an intricate system of associated arches and radial diaphragm walls to maintain rigidity, the whole designed to carry the weight of the dome over the voids between the piers. (4) The dome proper, a true coffered hemisphere on the inside. (5) A system of seven stepped rings and an upward extension of the cylindrical wall on the outside of the dome, these deviations from sphericity occupying the lower half of the exterior surface of the hemisphere. (6) The whole concrete fabric contains a graded aggregate, ranging from travertine at the base to pumice at the oculus, corresponding to the continuously decreasing compressive load from bottom to top.

These technical intricacies pass far beyond the pragmatic approach of the craftsman and reveal an understanding of structural mechanics that must have taken a fairly exact pictorial-geometric form in the mind of the builder.

The first three elements—foundation, drum, internal vaulting— carry the dead load of the dome to the eight radial piers and hence to the earth. The stepped rings and the upward extending walls of the drum sustain the horizontal component of the dome's radial thrust and absorb the tensile stress within the dome and the similar stress arising from the bending moment in any element of the cylinder. Both the dome and the cylinder, by virtue of the two-way curvature of the former and the single curvature of the latter, possess an inherent strength beyond that of their material mass to resist the forces acting on them. They are, for this reason, pioneer form-resistant structures.

Yet the scientific theory of the stress distribution and elastic behavior of domes was not developed until the twentieth century, and the associated structural form itself, the concrete dome, did not appear until 1900. We have a lag of nearly two thousand years between the art and its technical embodiment on the one hand and the scientific understanding on the other.

But more than this is involved, and it brings me back to our essential theme. Structures like the Pantheon reveal nature in two ways: first, they provide the symbolic restatement of the form of the divine cosmos or the natural order; second, they constitute a revelation of physical nature itself, through the seeming paradox of immense weight (5,500 tons of concrete and brick for the Pantheon dome) acting as though it were weightless. This hidden, even mysterious, capacity to span broad enclosures without apparent support, powerfully suggested the spherical canopy of the heavens, with the invisible supports of the undeviating celestial bodies.

The resources of experimental science, however, ultimately brought the technology of building design to the predictive stage, that is, to the stage where the action of the whole finished fabric could be laid out in prior algebraic and geometric symbols and built accordingly with full assurance that it would work as predicted. But a great many developments in science and mathematics had to be brought to maturity before this stage could be reached.

Building art provides the leading—perhaps now the only—example of techniques in which technology and its underlying scientific theory are still the handmaids of art. The supreme example is the Eiffel Tower, perhaps the only perfectly free structure, erected simply as a monument to itself. A good contemporary United States example is the Dulles Airport Terminal in Washington, where the technical solutions were dictated by aesthetic aims—chiefly sweeping space, buoyancy, and the drama of interacting structural elements. At the same time, the technology of uni-directional suspended systems was powerfully stimulated by serving these strictly formal ends.

Since there is no longer public agreement on a cosmos which is susceptible of translation into a symbolic image, the artistic-utilitarian object, like the airport terminal, is less a symbol than an expression of psychological states of which the prime ingredient, perhaps, is an evocation of kinaesthetic images.

Stages in the Development of Newton's Dynamics

By Richard S. Westfall, Indiana University

"IT IS A CLICHÉ that scientists do their best work when they are young," a recent article affirms. "This is not to say that scientists over forty are completely incapable of fruitful work, but most of the major scientific discoveries—especially in the fields that require the greatest gifts of abstract reasoning, such as pure mathematics and theoretical physics—are made by men in their twenties and thirties. Newton discovered the law of universal gravitation when he was twenty-four. . . . By the time he was forty, his truly creative days were over, and at the age of fifty he had given up mathematics completely."[1] Perhaps civilization, or at least the culture internal to the scientific community, has been accelerating since the seventeenth century; at least when I compare the assertion above with the facts of the scientific revolution, the cliché dissolves away into outright falsehood. It is not true with respect to Kepler; it is not true with respect to Galileo; it is not true with respect to Descartes; it is not true with respect to Huygens; it is not true with respect to Leibniz. Above all, for my purposes today, it is not true with respect to the man specifically mentioned.

Leaving aside the question whether Newton can be said to have discovered the law of universal gravitation in 1666, I want to focus my attention on a period of approximately six months at the end of 1684 and the beginning of 1685 when, at the age of forty-two, Newton began to compose the *Principia*. It is a fact that until this period Newton did not command a dynamics either consistent within itself or adequate to the task of expressing the *Principia*. Although he built on foundations he had prepared twenty years before, the formulation of his dynamics in a system capable of performing the tasks he set for it was the work of this period. If that was not an achievement in theoretical physics, I find it impossible to imagine what is.

The first stage in the development of Newton's dynamics belongs to the year 1664 when he composed a brief essay on "violent Motion" for his undergraduate notebook. In the essay, he considered three alternatives for the continuation of violent, that is, projectile motion after the projectile has separated from the projector, the classic problem to which the new conception of motion addressed itself. At some length, he rejected the theory that air can act as the mover, dealing not with the Aristotelian theory but with its vulgarized version, antiperistasis. Second, he considered whether the motion can be sustained by "a force imprest." The discussion is curious in that he took impressed force to mean the continued action of the original motor at a distance, and he refuted the theory by demonstrating the impossibility of such a force being communicated by either material or immaterial means. Hence he concluded for the third alternative, that a projectile is moved by its "naturall gravity."[2] The concept of gravity as an internal motive principle derived from the atomist tradition, which greatly influenced Newton's early philosophy of nature. For my purpose, it is significant that his effective introduction to the science of mechanics treated uniform motion as the product of a force internal to the moving body.

On January 20, 1665, Newton undertook an extensive examination of questions of motion and impact, which he recorded in the *Waste Book* as a set of definitions, axioms, and propositions entitled "Of Reflections." Quite a different idea of motion animated these notes. Whereas the atomist tradition stood behind the essay "Of violent Motion," the form in which Newton cast "Of Reflections" bears unmistakable marks of the influence of Descartes. After a couple of false starts, he put down eleven definitions of such things as motion and quantity of motion, followed by a series of axioms:

1 If a quantity once move it will never rest unless hindered by some externall caus.
2 A quantity will always move on in ye same streight line (not changing ye determination nor celerity of its motion) unless some externall cause divert it.[3]

Only an explicit reference to the state of rest would have been required to convert the two axioms into an affirmation of inertia indistinguishable from that in the first law of motion.

As the title, "Of Reflections," indicates, Newton's purpose was to solve the problem of impact with which Descartes had grappled unsuccessfully.

If 2 equall bodys (bcpq & r) meete one another wth equall celerity (unless they could pass through one y^e other by penetration of dimensions) they must mutually (since y^e one hath noe advantage over y^e other they must) equally hinder y^e one y^e others perseverance in its state. likewise if y^e body aocb be $=$ & equivelox wth r, they meeting would equally hinder or oppose y^e one y^e others progression or perseverance in their states therefore y^e power of y^e body aopq [aocb $+$ bcpq] (when tis equivelox wth r) is double to y^e power y^t r hath to persever in its state. y^t is y^t efficacy force or power of y^e caus wch can reduce aopq to rest must bee double to y^e power & efficacy of y^e cause wch can reduce r to rest, or y^e power wch can move y^e one must be double to ye power wch can move ye other soe y^t they be made equivelox. Hence in equivelox bodys ye power of persevering in their states are proportionall to their quantitys.[4]

Newton's approach to impact seized on exactly that aspect of the Cartesian approach which Huygens sought constantly to avoid. Newton's treatment, from its inception, was frankly dynamic. Already he had defined quantity of motion as the product of a quantity (that is, the quantity of matter in a body) and its velocity. Basic to his dynamics of impact was to be a unit of power or force defined in terms of quantity of motion. As he said in an earlier version of the passage above, "the motion of one quantity to another is as their powers to persever in that state."[5]

A concept of force measured in terms of quantity of motion was hardly a novelty to seventeenth-century mechanics. Quite the contrary. Indeed the history of mechanics in the seventeenth century offers a cautionary tale to students of scientific revolutions and paradigm shifts, suggesting that the flash of insight sometimes illuminates a pattern more cluttered with the rejected tradition than one might wish.

No two men contributed more to the new idea of motion on which the modern science of mechanics was built than Galileo and Descartes. From the vantage point of the twentieth century, the new idea of motion, which we can identify for convenience with the principle of inertia, implies a dynamics devoted to changes of motion. So obvious is the implication that more than one historian of science has labeled Galileo as the father of modern dynamics. This is a mistake. Not only did Galileo fail to develop a generalized dynamics which treated all changes of motion from a unified point of view, but the concept of force that came spontaneously to his mind was exactly the one that had no right to be there at all, a concept essentially identical with the rejected idea of impetus. For example, (and an indefinite number of analogous passages could be cited) he asserted that the "*impeto*" which the bob of a pendulum acquires in descending with a natural motion is able "to drive it upward by a forced motion" (*sospignere di moto violento*) through an equal ascent.[6] Similarly, Descartes' primary dynamic term was the "force of a body's motion." The unique position of the *Waste Book* in the history of dynamics derives from its recognition that a dynamics built on the principle of inertia demands a different concept of force. He realized that the "force of a body's motion" can be seen from another perspective. In impact, the force of one body's motion functions in relation to the second body as the "external cause" mentioned in axioms 1 and 2 as the sole means that can alter its state of motion.

Newton made the perspective of the second body his primary one. Thus, he was the first man fully to comprehend the implication of inertia for dynamics, that the prime necessity of an operative dynamics was a conceptual unit to measure the "external cause" of changes of motion. "Of Reflections" set out to convert the available idea of force to that use. After the initial set of axioms defining inertial states, he added two more that began the definition of force.

3 There is exactly required so much & noe more force to reduce a body to rest as there was to put it upon motion: et e contra.
4 Soe much force as is required to destroy any quantity of motion in a body soe much is required to generate it; and soe much as is required to generate it soe much is alsoe required to destroy it.[7]

It followed then that when equal forces move unequal bodies, their velocities are inversely as their quantities, and when unequal forces move equal bodies, their velocities are proportional to the forces.

104 Hence it appeares how & why amongst bodys moved some require a more potent or efficacious cause others a lesse to hinder or helpe their velocity. And ye power of this cause is usually called force.[8]

In "Of Reflections," Newton began to develop an abstract conception of force, separated from its cause and treated solely as a quantitative concept able to enter into a quantitative mechanics. In a context set by the introductory assertion of the principle of inertia and by the problem of impact, there was apparently only one quantity that could be the measure of force—change in motion. It is obvious, of course, that change of motion is not identical to rate of change of motion. Newton was seeking to quantify the phenomena of impact; total change of motion was the quantity that presented itself. Thus he measured force, not by ma or $(d/dt)mv$, but by Δmv.

A generation earlier, Descartes had established the ultimate foundation of the new mechanics by setting all motion, as motion, on the same place. To this principle Newton now added a second, that all changes of motion are equivalent and measurable on a linear scale. "Tis knowne by the light of nature," he asserted, "that equall forces shall effect an equall change in equall bodies."[9] As much force as is necessary to generate a given motion, so much exactly is needed to destroy it. "For in loosing or to [sic] getting ye same quantity of motion a body suffers ye same quantity of mutaion [sic] in its state, & in ye same body equall forces will effect a equall change."[10] In a body already in motion, an equal force will generate another increment of motion equal to the first, "since tis noe greater change for (a) to acquire another part of motion now it hath one y^n for it to acquire y^t one when it had none . . ."[11] He summed up the definition of force in axiom 23: "If ye body bace acquire ye motion q by ye force d & ye body f ye motion p by ye force g. yn d:q :: g:p."[12]

The concept of force that dominated dynamics in the seventeenth century before Newton was a legacy from impetus mechanics and the common-sense perceptions which it formalized. It was a legacy ultimately beyond reconciliation with the principle of inertia that was basic to the new science of motion.

Seen with the vantage of hindsight, the idea of inertia appears to have demanded a concept of force measuring, not motion, but change of motion, such as Newton proposed in the *Waste Book*. Did his formulation of such a concept signify his rejection of the point of view recorded in the essay "Of violent Motion," and his definitive endorsement of the principle of inertia which, in its statement as the first law of motion, was to function as the cornerstone of his mature mechanics? Again the flash of insight appears to have cast a flickering light. The *Waste Book* suggests that rather than seeing the one concept of force as the denial of the other, Newton sought to reconcile them in a unitary dynamics:

The force wch ye body (a) hath to preserve it selfe in its state shall bee equall to ye force w^{ch} [pu]t it into y^t state; not greater for there can be nothing in ye effect wch was not in ye cause nor lesse for since ye cause only looseth its force onely by communicaeting it to its effect there is no reason why its [sic] should not be in ye effect w^n tis lost in ye cause.[13]

Pursued to its ultimate implication, the passage seems to dissolve kinematics in a universal science of dynamics, in which the force of a body at any moment, expressed in its motion, is the sum of the forces that have acted upon it.

Would it be possible to reconcile such a dynamics with the princi-

ple of inertia on which the entire passage attempted to base itself?
Newton's statement of inertia in the *Waste Book* derived immediately
from Descartes', and for all his talk of the "force of a body's motion,"
Descartes had drawn from the principle of inertia the conclusion of
the relativity of motion, a conclusion impossible to reconcile with the
universal dynamics toward which Newton was groping. Moreover,
the uniformly linear relation of motion and force that he adopted,
whereby no less force is required to add a second increment of motion
to a moving body than to start it from rest and give it a first increment,
implied the dynamic identity of uniform motion and rest. Apparently
Newton perceived the incompatibility as well. At least he was
uneasy enough with the passage above to strike it out, and to
replace it with another in which he seemed deliberately to abandon
the idea that the force of a body's motion could be a meaningful absolute quantity in an inertial mechanics. Although he continued to
equate force with motion in the *Waste Book*, the "force of a body's
motion" largely ceased to connote a sustaining cause of uniform
motion and approached our concept of momentum. Its quantity
merely expressed the amount of force necessary to generate that
quantity of motion in an inertial frame of reference.[14]

The purpose of the entire exercise was the treatment of impact, and
exactly the treatment of impact served to emphasize a concept of
force as the measure of changes of motion instead of the measure of
motion itself. Examining what he called "ye mutual force in reflected
bodys," Newton asserted that "soe much as p presseth upon r so much
r presseth on p. And therefore they must both suffer an equall mutation in their motion."[15] Except for the special case in which p and r
have equal motions, the conclusion that they press each other equally
had seemed obviously false to everyone before who had attempted to
analyze the force of percussion. The *Waste Book* suggests that Newton came to deny the intuitively obvious by recognizing that for every
impact there is a frame of reference, that of the center of gravity, in
which the two bodies do have equal forces.[16] That is, his treatment
of impact depended on his accepting a relativity of motion in terms of
which the idea of an absolute force of a body's motion is meaningless.

The *Waste Book* also contains Newton's successful analysis of
circular motion in which he arrived at a quantitative statement of a
body's "conatus" or "endeavor" to recede from the center about
which it revolves. Whatever the success of the analysis, it embodied
another problem for his concept of force. Whereas impact presented
one model for the conceptualization of force, the endeavor to recede

from the center presented quite another. He suggested that the instantaneous endeavor to recede is "like the force of gravity . . ."[17] Here Galileo's analysis of free fall as uniformly accelerated motion was implicitly brought forward as a model for the conceptualization of force. Without evident hesitation, Newton applied the term "force" to the one as readily as to the other. The acceleration of a falling body, after all, is a change of motion quite as much as that which a body suffers in impact. Only in so far as he spoke of the "total force" exerted in a revolution did he recognize the dimensional incompatibility of the two cases. What then was the measure of force, ma or Δmv? Newton did not formulate the question, let alone answer it.

Newton's flirtation with the Cartesian idea of motion ended in disillusionment rather than marriage. Sometime not long after the end of his undergraduate career, he composed the violently anti-Cartesian *De gravitatione et aequipondio fluidorum*, in which, among other things, he returned to the atomist view of motion in order to escape from the terrors of Cartesian relativity. Physical and absolute motion, he asserted, "is to be defined from other considerations than translation, such translation being designated as merely external."[18] If translation is merely external, what can be essential? What except force? "Force is the causal principle of motion and rest. And it is either an external one that generates or destroys or otherwise changes impressed motion in some body; or it is an internal principle by which existing motion or rest is conserved in a body, and by which any being endeavours to continue in its state and opposes resistance."[19] It is true that the only examples he gave of forces that distinguish true from apparent motions were centrifugal forces. Nevertheless, when he said that "physical and absolute" motion is to be defined by something other than translation, Newton had more than circular motion in mind. The purpose of his argument, be it remembered, was to establish the existence, not of absolute space, but of absolute motion—absolute motion in refutation of relative motion, of which one could assert with assurance neither its velocity nor its determination. In circular motion, he found his readiest example, but he clearly intended it as no more than an example. Among his further definitions, that of inertia as "force within a body, lest its state should be easily changed by an external exciting force" could be reconciled with the relativity of motion and looked forward to Newton's mature mechanics. The definition of impetus, however, reasserted the concept of an internal force associated with uniform motion. "Impetus is force in so far as it is impressed on a thing."[20] He also defined gravity and conatus in terms of force. The basic message of *De gravitatione* is the displacement of relativistic kinematics by absolutistic dynamics, in which force rather than translation expresses ultimate reality.

With *De gravitatione*, Newton's early work in mechanics reached its terminus. During the following fifteen years, he put mechanics aside until the composition of *De motu* in 1684 announced a new period of intense activity from which the *Principia* emerged. When Newton picked up the thread again, he found the tangles of the 1660's still there. Above all, two major ambiguities associated with the concept of force remained. Is force the measure of motion or the measure of change of motion? If it is the latter, is its paradigm case impact or free fall, is it measured by Δmv or by ma? These questions remained to plague the composition of the *Principia*.

The original version of *De motu* began with the definition of two kinds of force, centripetal force and inherent force.[21] I am interested here in the second. "And [I call that] the force of a body or the force inherent in a body [*vim . . . corpori insitam*] by which it indeavors to persevere in its motion in a right line."[22] Lest the meaning of the definition should be unclear, he added an Hypothesis 2 which placed its meaning beyond doubt. "Every body by its inherent force alone proceeds uniformly in a right line to infinity unless something extrinsic hinders it."[23] Perhaps nothing illustrates more clearly the intent of *De gravitatione* when it announced that something other than mere translation is essential to motion. The nearest approach to the principle of inertia in *De motu* treated uniform rectilinear motion as the product of a uniform force, the *vis insita* of the moving body.

In the manuscript in which it has survived, the first version of *De motu* contains two stages, an original form and a set of emendations and additions. The original form started with the two definitions cited above and two hypotheses. When he revised the treatise, Newton added a third hypothesis in the margin together with the heading "Hyp 4," although he did not write in the fourth hypothesis itself. The third one made explicit the dynamics employed in the following propositions. "Hyp. 3. When it is acted upon by [two] forces simultaneously, a body is carried in a given time to that place to which it would be carried by the forces acting separately in succession during equal times."[24]

Newton had already employed the parallelogram of forces in Theorem 1. By the action of its inherent force alone, a body moves in a straight line, while impulses of centripetal force at equal intervals of time divert it from the original path into new ones. Using the parallelogram of forces and some simple geometry, Newton was able to demonstrate that Kepler's law of areas is satisfied under such conditions.[25] Thus *De motu* embodied the ideal of mechanics that *De gravitatione* had presented, and sought to submerge kinematics in a universal dynamics.

By returning to *De gravitatione*, Newton plunged the dynamics of *De motu* into a series of contradictions to the resolution of which the work of the next six months was largely devoted. The definitions, first two and later a third, posited three kinds of force—inherent force, centripetal force, and resistance. The parallelogram of force was introduced as the means to determine the result when two of the forces act conjointly. Could they be compared in this way? In fact, the word "force" (*vis*) was the only common factor in concepts that were otherwise utterly disparate. The inherent force of definition two was conceived to maintain a uniform rectilinear motion, the force introduced originally to evade Cartesian relativism. Centripetal force and resistance, on the other hand, repeated the concept of force developed in the *Waste Book*, that of forces external to a body which act to alter its uniform rectilinear motion.

Nothing illustrates the incompatibility of the two conceptions of force more clearly than the fourth hypothesis, if we may assume that the statement Halley entered into version two of *De motu* after he had copied the imperfect version one, was what Newton had in mind when he wrote "Hyp 4" in the margin. In Halley's version, the space that a body, "under the action of any centripetal force whatever, describes in the very beginning of motion is proportional to the square of the time."[26] In version three of *De motu*, Newton offered a demonstration of the proposition (now called Lemma 2) that is interesting for what it assumes. When we read the proposition, we are apt to think that its point is the proportionality of distance to the square of time, so that a uniform force is seen as the cause of a uniform acceleration. In fact, this is what Newton assumed—referring it to Galileo's exposition. The point lay, rather, with the word "whatever." He was concerned to demonstrate that with a nonuniform centripetal force, as well as with a uniform one, the distance described at the very beginning of motion is proportional to the square of the time.[27]

Whereas "force" as inherent force causes a uniform motion ($F = mv$), "force" as centripetal force causes a uniform acceleration ($F = ma$). Newton's parallelogram of forces was an adaptation of Galileo's parallelogram of motions, an adaptation which expressed his principle that force is more basic than motion. The two forces in question here, however, have utterly different relations to motion.

Newton's parallelogram assumed that they are identical in their relations to motion.[28]

Nor was this the only problem that beset the conception of force in *De motu*. As far as inherent force is concerned, we can overlook the minor ambiguity arising from the fact that *De motu* does not contain a definition of mass and say that inherent force is measurable by *mv*. What is the measure of external forces which alter uniform motion? In so far as *De motu* defined them, it appears that such forces are measured by *ma*. Hypothesis 4 explicitly assumed that a uniform (centripetal) force produces a uniform acceleration, and Theorems 2 and 3, to name no more, adopted the same measure of force. Theorem 1, on the other hand, employed a force that acts at equal intervals "with a single but great impulse," generating a corresponding motion—a return to the concept of the *Waste Book* in which force is measured by Δmv.[29]

Version two of *De motu*, which was little more than a fair copy of the amended form of version one, contributed nothing to the development of Newton's dynamics. Version three, in contrast, marked a further stage. Hypotheses 3 and 4 of the earlier versions were moved to the status of Lemmas, enhanced by demonstrations, but not otherwise altered. To the remaining two hypotheses Newton added three more, decided that the word "hypothesis" did not express his true intent, struck it out in all five cases, and replaced it with a new word, "*Lex*."

In their original incarnation, then, the laws of motion numbered five. The first remained substantially unchanged from version one, still asserting that by its inherent force alone a body proceeds in a right line. Law two, however, underwent a significant change. The word "impressed" replaced the word "centripetal," not only increasing the generality of the law, but also expressing the dichotomy that Newton sought to exploit. The inherent or internal force of a body maintains it in uniform motion. The force external to a body, impressed on it from without, alters its uniform motion: "The change of motion is proportional to the impressed force and is made in the direction of the right line in which that force is impressed."[30] With the exception of one word (the adjective "motive" modifying impressed force), the law is identical with that which appeared in the *Principia*. Despite the fact that Lemma 2 and divers theorems continued to assert a conception of force that can only be measured by acceleration, Law 2 embodied the definition of force in terms of Δmv.

Of the remaining three hypotheses—or laws—added in this version, the last of them formalized what he had discovered in version one about the resistance of a medium. While it was important in the history of mechanics, it was certainly not a law of motion, as Newton recognized when he later demoted it from the list.

The other two laws of version three were potentially more troublesome. One asserted that the "relative motions of bodies contained in a given space are the same whether the space in question rests or moves perpetually and uniformly in a straight line without circular motion." The other complemented it by affirming that the common center of gravity of a system of bodies does not change its state of motion or rest because of the mutual actions of the bodies on each other.[31] These two laws returned to the insights of the *Waste Book* and its realization that in an inertial system two bodies isolated from outside influences can be considered as one concentrated at their center of gravity. Thus he referred the second of them to Law 2, the force law, for confirmation. Law 2 itself implied the concept of inertia in its renewed assertion of the linearity of all changes of motion.

In version three of *De motu*, then, three new laws of motion marked the beginning of Newton's return to inertial mechanics. Law 1, on the other hand, contradicted them directly in its assertion that uniform motion is distinguishable by the internal force that sustains it. Newton also inserted a Scholium after Problem 5 in which he referred to "the immense and truly immobile space of the heavens . . ."[32] Thus two contradictory currents of thought appear to have governed the amendments that went into version three, as though Newton felt the need to assert the reality of absolute space the more he undermined its operative significance as he clarified his dynamics.

In the papers revising version three that stood between it and the so-called *Lectiones de motu*, the divergent tendencies become fully manifest. A greatly expanded set of definitions began with absolute time, relative time, absolute space, and relative space. After definitions of body, center and axis of a body, place, and rest, Newton continued with a definition of motion which came immediately to grips with his central concern, the distinction of absolute from relative motion. In circular motions, the endeavor to recede from the center enables us to determine an absolute rotation. In general, the distinguishing factor is force.

> Furthermore, that motion and rest absolutely speaking do not depend on the situation and relation of bodies between themselves is evident from the fact that these are never changed except by a force im-

pressed on the body moved or at rest, and by such a force, however, are always changed; but relative motion and rest can be changed by a force impressed only on the other bodies to which the motion and rest are related and is not changed by a force impressed on both such that their relative situation is preserved.[33]

Both in their form and in their content, the definitions seem to return to the essay *De gravitatione* with its impassioned rejection of Cartesian relativity and its conviction that relativity can be escaped via dynamics. The dichotomy of absolute and relative motion was equally the dichotomy of kinematics and dynamics.

And exactly here was the dilemma. Farther down the page, the definition of the inherent force of a body, that which is not merely an external relation—what in this paper, as though consciously recalling the earlier assertion, he first called the "force . . . inherent and innate in a body . . ." (*Vis corporis seu corpori insita et innata . . .*), and then the "inherent, innate and essential force of a body . . ." (*Corporis vis insita, innata et essentialis . . .*)[34]—was in process of a revision which would render it useless in the determination of absolute motion. In its first three versions, *De motu* had defined the inherent force of a body as that by which it endeavors to persevere in uniform motion in a right line, and the first law of motion had reaffirmed what the definition asserted. "The inherent, innate and essential force of a body," he now stated, "is the power by which it perseveres in its state of resting or of moving uniformly in a right line, and is proportional to the quantity of the body; in fact it is exerted proportionally to the change of state."[35] The first law of motion underwent a corresponding change, in effect abandoning the concept that a uniform motion is the product of a uniform force, converting inherent force into a resistance to change in a body's state, and tacitly embracing the principle of inertia.

In light of the direction in which his dynamics was tending, we cannot avoid asking whether Newton's insistence on absolute motion had any practical consequences for his mechanics. Not only do I fail to find any such consequences, not only do I fail to find in Newton any criterion by which an absolute motion (as opposed to a narrowly confined set of absolute rotations) might be identified, but I do find the burgeoning growth of Newton's discussion of absolute motion to be revealing. In the process by which *De motu* expanded into the *Principia*, the space devoted to the argument for absolute motion grew in size with every change in Newton's dynamics that diminished its operational relevance. Newton's assertion of absolute motion has all the appearance of an act of defiance hurled in the face of the very current of thought on which his dynamics itself was borne inexorably toward its ultimate form.

The changes introduced in the revisions of version three of *De motu* extended beyond the rejection of his original notion of inherent force. It was in this manuscript that Newton consciously moved outside the limited range of celestial dynamics and attempted systematically to examine the basic concepts of dynamics as a whole. In its original form, *De motu* had been far more confined. To solve the mechanics of planetary motion, Newton had seized on a minimal number of dynamic concepts ready at hand but unexamined. The original version contained two definitions only, and version three contained only four. At one time or another as it underwent continual emendation, the revision of version three proposed eighteen definitions of quantities possibly required for a systematic dynamics. With the redaction of this paper, Newton's dynamics assumed a consciously generalized formulation and began to approach its ultimate shape.

One of the quantities that struggled toward precise definition was quantity of matter or mass. In reality the definition of quantity of matter furnished only half of Newton's concept of mass. The other half came from his revised idea of inherent force.

> The inherent, innate, and essential force of a body is the power by which it perseveres in its state of resting or of moving uniformly in a right line, and is proportional to the quantity of the body; in fact it is exerted proportionally to the change of state, and in so far as it is exerted, it can be called the exerted force of a body.[36]

As he thought about it, Newton liked the idea of exerted force and added a definition of it. Exerted force is that by which a body strives to preserve "that part of its state of moving or of resting which it loses in single moments and is proportional to the change of that state or to the part lost in single moments."[37] Significantly, in a later revision of these definitions Newton altered that of inherent force so that it read, not "inherent force of a body," but "inherent force of matter," as it remained in the *Principia*. In any given body, he continued, the inherent force of matter is proportional to the mass and differs in no way from the inactivity of the mass.[38]

Contrary to Newton's assertion, the inactivity of matter as conceived by the prevalent mechanical philosophy was exactly what inherent force, as he now defined it, did differ from. If matter is endowed with an inherent force by which it resists efforts to change its state, it cannot be wholly inactive, or in the classic phrase of the seventeenth century, wholly indifferent to motion. Indeed, Newton's formulation of the laws of mechanics implied a conception of matter at once inert and active, unable to initiate any action itself, passively dominated by external forces, but endowed with a power to resist their actions. Through the *Principia* this strange conception of matter passed into the mainstream of modern science, where familiarity coupled with the power of the dynamics that employs it has dulled our perception to its ultimate paradox. In the *Principia*, Newton himself summarized the paradox in another anomalous phrase, *vis inertiae*, which we might translate freely as "the activity of inactivity," or perhaps "the ertness of inertness."

Given the other factors already established in the science of mechanics, it is difficult to see how a workable dynamics could have developed without some such idea of matter. The indifference of matter to motion expressed admirably Galileo's basic insight that a body in motion will continue in motion. Meanwhile it was apparent to everyone that matter cannot be wholly indifferent to motion since unequal amounts of effort are required to cause equal changes of velocity in unequal bodies. Struggling to express this observation, Descartes had said that a body persists, as much as in it lies, in its motion in a straight line. With its connotations of activity applied to inert matter, the verb "persist" foreshadowed the paradox of Newton's *vis inertiae*.

Similarly, the analysis of impact in the *Waste Book* had stated as a basic proposition that equal forces applied to unequal bodies generate motions inversely proportional to the sizes of the bodies. Facing the same problem of impact, Mariotte was to arrive at the same idea of an internal resistance; and in the century's most consciously analytical examination of the question, Leibniz concluded that if matter were indifferent to motion, then any force would be able to generate any motion in any body. Not unlike Newton, Leibniz reformulated the conception of matter and defined its essence as force rather than extension. Newton did not go that far. To him, *vis inertiae* represented one of the universal properties of matter, not displacing extension, but standing equally beside it together with hardness and impenetrability. Inevitably, he set it proportional to the quantity of matter. As he revised *De motu* into the *Principia*, he returned specifically to Descartes' discussion of motion which had influenced the passage in the *Waste Book*. The inherent force of matter is a power of resisting by which "any body perseveres, as much as in it lies, in its state of resting or of moving uniformly in a right line . . ."[39] Because of the inherent force of matter, there is a constant proportion between the force applied and the change it generates in a body's state of motion. Without some such concept, a quantitative dynamics would have been impossible.

Newton devoted part of the re-examination of basic terms that contributed to the expansion of the definitions in *De motu* to the concept of force itself. *De motu* had started with only inherent force and centripetal force, to which he added resistance not much later. In the revision of version three he now defined six kinds of force—inherent force, the force of motion, exerted force, impressed force, centripetal force, and resistance. "There are also other forces," he added, "arising from the elasticity, softness, tenacity, etc. of bodies, which I do not consider here."[40]

I have already discussed inherent force, in the revised form in which it now appeared. The force of motion represented a last effort to save the idea that had been eliminated from the revised definition of inherent force. Exerted force, on the other hand, supplemented the new concept of inherent force. Whereas inherent force is proportional to mass, exerted force is proportional to the impressed force and thus to the change of the body's state of motion or rest, and is not improperly called the reluctance or resistance of the body. One species of exerted force is the centrifugal force of revolving bodies.[41] Obviously the idea of exerted force was a groping step toward the insight expressed in the third law, first announced later in this same paper. If a body resists the actions seeking to change its state of motion, it exerts in reaction a force on whatever acts on it.

The other three forces, impressed force, centripetal force, and resistance, were in fact identical in Newton's opinion; centripetal force and resistance were merely specific forms of impressed force. The phrase "impressed force" had appeared in the wording of the second law of motion in version three of *De motu*, although the definitions, which included centripetal force and resistance, had not included it. Now he defined impressed force as a general term subsuming the other forces that act on a body from outside to alter its state of motion or rest.

> The force brought to bear or impressed on a body is that by which the body is urged to change its state of moving or resting and is of divers kinds such as the pulse or pressure of percussion, continuous pressure, centripetal force, resistance of a medium, etc.[42]

When Newton proceeded to cancel his definitions of the force of motion and exerted force, two kinds of force remained as the foundation of his dynamics—the inherent force of matter, by which a body resists efforts to change its state of motion or rest, and impressed force, any action arising outside a body which attempts to change its state of motion or rest. Inherent force, which is proportional to the quantity of matter or the mass, establishes a constant proportion between an impressed force and the change of motion it produces.

The suggestion has been advanced that Newton's use of the adjective "impressed" was consciously intended to distinguish mere "force" from "impressed force." The point of the suggestion rests on the specific form in which Newton couched Law II: "The change of motion is proportional to the motive force impressed . . . ."[43] We are accustomed to word the second law of motion somewhat differently so that it refers, not to change of motion (Δmv), but to rate of change of motion (ma). The distinction between force and impressed force proposes to resolve this dilemma by the argument that Newton understood "impressed force" to mean the application of force over a period of time ($\int F dt$).[44] I am convinced that the dilemma has no resolution in Newton's works, that the problem visible in the *Waste Book* remained a problem unsolved, and that he used the term "force" indiscriminantly both in contexts where it meant Δmv and in others where it meant ma. In this, Newton repeated the dimensional confusion of seventeenth-century mechanics as it strove to define useful dynamic concepts.

However, in Newton's case the consequences of the confusion were virtually eliminated. When he used "force" in the sense of Δmv, as in Theorem 1, he defined the problem so that separate impulses follow each other at equal periods. If time is not in the definition of force, it is in the definition of the problem. In the passage to the limit, Newton understood the total quantity of force in the impulse to be, as it were, spread out over the interval of time—a procedure that cannot bear dimensional analysis, but one that led to no confusion in practice. Meanwhile, the development of the definitions in *De motu* seems clearly to argue that he intended no distinction between force and impressed force. The fact that he built a dynamics on divers concepts of force but not on force unmodified, which he never undertook to define, is surely decisive. The distinction is not between "force" and "impressed force," but rather between "inherent force" and "impressed force," *vis insita* and *vis impressa*.

The distinction was expressed most aptly by the contrasting adjectives "internal" and "external," which he used in *De gravitatione*. The extra adjectives that he added in the revision of verse three, even if they later dropped out of sight, also help to illuminate the distinction—on the one hand "the inherent, innate, and essential force of a body" (*corporis vis insita, innata et essentialis*), and on the other hand "the force brought to bear and impressed on a body" (*vis corpori illata et impressa*). Although the revisions of version three of *De motu* did not include a statement of the principle of inertia, they implied it, and they set the enduring pattern of Newton's dynamics— the establishment of equilibrium between inherent and impressed forces by means of changes in inertial motion.

One cannot mistake the central role of the model of impact in his ultimate vision of dynamics. On the one hand, it furnished his definition of impressed force, although I have argued that Newton understood the definition not to stand in conflict with a definition drawn from the model of free fall. More important, impact supplied the insight formalized in Law III in a way that free fall could not have done. And Law III, in establishing the relations of inherent and impressed force, provided the capstone of his dynamics.

In every essential way, the revisions of version three of *De motu*, which apparently took place early in 1685, fixed the elements of Newton's dynamics in their permanent form. The definitions themselves underwent one further revision that set their wording substantially as the *Principia* published them. In the first revision, the definitions culminated in six laws of motion. The sixth, a statement of the resistance of media, had the dignity of law stripped from it by a stroke of the pen. The *Principia* moved Laws 4 and 5 to the status of corollaries to the laws of motion. The first three are recognizably the three laws of motion as we know them, requiring merely improved wording (and in the case of Law 2, little even of that) to attain their permanent form. Only Law 1 appears strange with its continued reference to *vis insita*. The redefinition of *vis insita*, however, had left it redundant in the first law, as Newton recognized in the so-called *Lectiones de motu*. With the mere elimination of the phrase, Law 1 emerged as the classic statement of the principle of inertia.

When the scope of version three of *De motu* is compared with the *Lectiones de motu*, composed initially later in 1685, the importance of the revisions becomes manifest. With the inherently unsatisfactory dynamics of *De motu*, Newton had solved the central problems of orbital dynamics and had worked out his basic approach to motion through resisting media. In neither case could his principles have withstood systematic criticism, and it is difficult to believe that they could have supported the immensely expanded investigation into mechanics celestial and mundane that Newton launched in 1685 and recorded in the *Lectiones*. With the new dynamics formulated in the revisions, however, far more was possible. Indeed, the *Principia* and the law of universal gravitation were possible—a satisfactory return, perhaps, on the investment of effort.

1. Jeremy Bernstein, *"Apologia pro Aetas Sua,"* (a review of G. H. Hardy's *Mathematician's Apology*), *The New Yorker* (February 1, 1969), 80.

2. University Library, Cambridge, *Add. MS. 3996*, ff. 98, 98v, 113–14. The passage is printed in John Herivel, *The Background to Newton's* Principia, (Oxford, 1965), 121–25.

3. *Ibid.*, 141. Herivel has published all of the material on mechanics in the *Waste Book*.

4. *Ibid.*, 153–54.

5. *Ibid.*, 137.

6. Galileo Galilei, *Dialogue Concerning the Two Chief World Systems*, tr. Stillman Drake, (Berkeley and Los Angeles, 1962), 227. I have checked the English of this and of similar passages against the original Italian.

7. Herivel, *Background*, 141.

8. *Ibid.*, 156.

9. *Ibid.*, 157.

10. *Ibid.*, 158.

11. *Ibid.*, 157.

12. *Ibid.*, 150.

13. University Library, Cambridge, *Add. MS. 4004*, f. 12v. Since Newton canceled this entry, Herivel has not published it.

14. Consider, for example, Axiom 118, which derived a general formula relating force, mass, and velocity. Given a body p moved by a force q and a body r moved by a force s, what is the ratio of their velocities, v and w? Newton demonstrated that

$$\frac{v}{w} = \frac{qr}{ps} \qquad (or \qquad \frac{{}^v p}{{}^v r} = \frac{{}^f p^m r}{{}^f r^m p} \quad)$$

to wit, that "y^e motion of p is to y^e motion of r as y^e force of p is to y^e force of r." In the following sentence, however, he insisted that he understood changes of motion to be in question: "And by y^e same reason if y^e motion of p & r bee hindered by y^e force q & s, y^e motion lost in p is to ye motion lost in r, as q is to s. or if ye motion of p be increased by ye force q, but ye motion of r hindered by ye force s; then as q, to s :: so is the increase of motion in p, to ye decrease of it in r." Herivel, *Background*, p. 159.

15. *Ibid.*, 159.

16. A series of propositions established that two bodies have equal motions in relation to their common center of gravity, and that the center of gravity of two bodies in uniform motion is also in uniform motion, both when the two bodies are in the same plane and when they are in different planes. Newton was now ready to demonstrate that their center of gravity also continues to move uniformly when the two bodies meet in impact and rebound.

For ye motion of b towards d ye center of their motion is equall to ye motion of c towards d . . . therefore ye bodys b & c have equall motion towards ye points k & m y^t is towards ye line kp [the line of motion of their center of gravity before impact]. And at their reflection so much as (c) presseth (b) from ye line kp; so much (b) presseth (c) from

it . . . Therefore e & g [the locations of b and c at some time after im-
pact] have equall motions from ye point 0. w^{ch} . . must therefore be y^e
center of motion of ye bodys b & c when they are in ye places g & e,
& it is in ye line kp.
Ibid., 168–69.

17. University Library, Cambridge, *Add MS. 4004*, f. 1. Herivel has not
published the paragraph in which this phrase occurs along with the other
material from f.1 that he has published.

18. A.R. Hall and Marie Boas Hall, *Unpublished Scientific Papers of Isaac
Newton*, (Cambridge, 1962), 128.

19. *Ibid.*, 148.

20. *Ibid.*, 148.

21. The period of approximately six months at the end of 1684 and the
beginning of 1685 that I have referred to was the time in which Newton was
composing and revising *De motu*. The dates assigned are uncertain but not
wholly conjectural.

Three distinct versions of *De motu* exist (the first entitled *De motu
corporum in gyrum*, the second bearing no title, the third *De motu sphaeri-
corum corporum in fluidis*), together with two other sets of papers that are
revisions of the third version (the first entitled *De motu corporum in mediis
regulariter cedentibus* and the second merely *De motu corporum*), preparatory for the so-called *Lectiones de motu*. The *Lectiones* themselves, which
were certainly not prepared as lectures but rather as a draft of Book I of the
Principia, went through at least one major revision which can be traced in
the manuscript.

An extensive literature on *De motu* exists, attempting among other things
to identify which (if any) of the three versions was the paper Edward Paget
carried to London in November, 1684, presumably to fulfill Newton's pledge
to Halley in August. To me, at least, it appears that this can only have been
version one (as published in Herivel, pp. 257–74). As it was originally written,
version one had two "Hypotheses." In a revision, Newton added a third in the
margin and entered a heading "Hyp 4" with nothing after it. (See the reproduction in Herivel, Plate 5 between p. 292 and p. 293.)

Version two (identical with that in the register of the Royal Society as
published by Rouse Ball—I have not seen the original—and also identical,
it is said, with a copy in the Macclesfield Collection that I have not seen) is
in Halley's hand. When it was copied, a heading for Hypothesis 4 was entered and a space for it was left, but the hypothesis itself was added later in a
different ink. I can only explain this by assuming that Halley copied Paget's
paper, which letters testify to have been in great demand, and carried it with
him on his trip to Cambridge before December 10. There he entered the
hypothesis in the place left for it, as well as some other material that also
appears in the distinct ink. The completed paper must then have been copied
into the register of the Royal Society and dated December 10, the day Halley
reported the paper to the Society.

Version three contains an interesting amendment; a reference to the
satellites of Saturn is crossed out. A letter from Flamsteed on December 27
questioned their existence; Newton's letter of December 30 asked specifically
about them. On this evidence, I am prepared to say that version three was
composed around the end of December. In a paragraph added in the third

version to the end of Theorem 4, Newton raised the question of the mutual action of the planets on each other, saying that he was not then in a position to deal with it. The paragraph at least recalls his correspondence with Flamsteed at that same time (see his letters of December 30, 1684, and January 12, 1684–85).

I know of no direct evidence by which to date the two papers revising version three, but I assume they were composed early in 1685. By internal evidence they preceded the so-called *Lectiones de motu*. The *Lectiones* in turn are not dated. Since they constitute in reality a considerably expanded version of *De motu*, and since Newton's correspondence with Halley places their revision in the winter of 1685–86, there is some foundation for placing their composition in the first half of 1685. The dates later added to them, suggesting that they had been delivered as lectures beginning in October, 1684, have no meaning. Newton was accustomed to fulfill the obligation to deposit his lectures by sending over any handy papers at a later time. Other supposed lectures have been shown to differ from what was delivered, and there is no reason to think these papers, which have no semblance to lectures but every semblance to an intended book, are different. In fact, the lecture numbers, added in the margins, run consecutively over a first draft and a later revision of the same material. It appears to me that the conjectural period of six months late in 1684 and early in 1685 has considerable evidence to support it.

22. Herivel, *Background*, p. 257 and p. 277. I am citing all of the material in *De motu* from Herivel, who has published all of the versions. I give both the location of the original Latin and the location of his English translation, from which mine frequently differs. Since I prepared translations of my own, I prefer to use them, but I have consulted his and drawn on them to a considerable extent.

23. *Ibid.*, 258, 277.

24. *Ibid.*, 258, 278.

25. *Ibid.*, 258, 278. This theorem together with a couple of others remained substantially unchanged through revisions of *De motu*, and hence propositions in Section I of Book I of the *Principia* contain phrases about a body moving in a straight line *sola vi insita*.

26. *Ibid.*, 258, 278.

27. *Ibid.*, 295–96, 300. Cf. Theorem 3 of *De motu*, corresponding to Proposition VI of the *Principia*. It establishes a general geometric expression for the centripetal force necessary to generate any curve. The line PR is tangent to the curve at point P; from a neighboring point Q on the curve, the line QR parallel to the radius vector to P, joins Q to the tangent. The line QR, Newton stated, "varies with the centripetal force when the time is given and with the square of the time when the force is given, and hence, when neither is given, varies with the centripetal force and the square of the time together." *Ibid.*, 260, 279–80.

28. Although version one had no diagram of the parallelogram, version three did. The wording of what was now called Lemma 2 had also changed. "When it is acted upon by [two] forces simultaneously, a body describes the diagonal of a parallelogram in the same time in which it would describe the sides under the action of the forces separately." (*Ibid.*, 295, 299.) In fact his use of the parallelogram in the propositions of version one assumed the same thing.

It may be remarked in passing that this version of Lemma 2 appeared, with only a minor change of wording, as Corollary I to the Laws of Motion in the first edition of the *Principia*. (*Corpus viribus conjunctis diagonalem parallelogrammi eodem tempore describere, quo latera separatis.*) Only in the second edition did he eliminate the paradox that such a statement presents as a Corollary to Law I, which states the principle of inertia. In fact he did not alter the wording of the Corollary, but where in its demonstration edition one referred to a body being "carried" from A to B by force M and from A to C by force N, edition two spoke of the forces being "impressed" on the body at point A. The parallelogram returned then to its original form, a parallelogram of motions, in this case of motions set proportional to the (total) forces which generated them. Even in its amended form it remains a curious artifact giving witness to the development of Newton's dynamics.

29. *Ibid.*, 258, 278.

30. University Library, Cambridge, *Add. MS. 3965.7*, f. 40. Although the wording is nearly identical to that in the *Principia*, Newton was not satisfied with it at the time. He changed the initial phrase to read "The motion generated or the change of motion . . . ," and then changed it a second time to the form it retained in version three: "The change in the state of moving or of resting is proportional to the impressed force . . . ." Herivel, *Background*, 294, 299.

31. *Ibid.*, 294, 299.

32. *Ibid.*, 289, 302.

33. *Ibid.*, 305, 310.

34. University Library, Cambridge, *Add. MS. 3965.5a*, f. 26; Herivel, *Background*, 306, 311.

35. *Ibid.*, 306, 311.

36. *Ibid.*, 306, 311.

37. University Library, Cambridge, *Add. MS. 3965.5a*, f. 25v. Herivel, publishes this definition with the second revision. *Background*, 317, 320.

38. *Ibid.*, 315, 318.

39. *Ibid.*, 315, 318. Cf., I. B. Cohen, " '*Quantum in se est*': Newton's Concept of Inertia in Relation to Descartes and Lucretius," *Notes and Records of the Royal Society of London*, 19 (1964), 131–55.

40. Herivel, *Background*, 306, 311.

41. University Library, Cambridge, *Add. MS. 3965.5a*, f. 25v. Cf. note 37.

42. Herivel, *Background*, 306, 311.

43. *Mathematical Principles of Natural Philosophy*, tr. Andrew Motte, revised by Florian Cajori, (Berkeley, 1960), 13.

44. R.G.A. Dolby, "A Note on Dijksterhius' Criticism of Newton's Axiomatization of Mechanics," *Isis*, 57 (1966), 108–15.

Commentary on the Paper of Richard Westfall

By Thomas L. Hankins, University of Washington

IN AN EARLIER PAPER by Professor Westfall on the problem of force in Galileo's physics I came across this warning:

> Obvious perils are involved when a Newtonian scholar is invited to a conference on Galileo. He is apt to confuse the two ends of the seventeenth century and to employ the achievement of Newton as a criterion by which to measure that of Galileo. Such proceedings would constitute a public scandal . . . the Newtonian scholar who dares to discuss Galileo in the presence of Giorgio de Santillana is obviously in greater danger than ever Galileo was.[1]

The same danger confronts anyone commenting on a paper by Professor Westfall dealing with Newton's dynamics. The time when one can philosophize on Newton's mechanics from his definitions and his three laws of motion alone is past. And with an awareness of Professor Westfall's thorough study of the Newton manuscripts, I comment with considerable caution.

It seems to me that the first thing Professor Westfall is telling us is that Newton's ideas in dynamics did not appear in an instant, that they matured over a long period of time, and that they were not the product of his early years. Newton was a genius, but we must avoid the romantic inclination to make all scientific geniuses superhuman as well. It is enough to recognize Newton's accomplishments without insisting that they all came before the magic age of forty.

Nor is it necessary to argue that the laws of motion in the *Principia* could not be improved upon, or that they represented a totally new departure from the mechanics of Descartes and Galileo. Professor Westfall has done us a favor by indicating how Newton groped his way through the murky concepts of force and mass to a more general and more usable mechanics. He has shown that philosophical and even theological speculation seemed to be as important to Newton as his vaunted experimental method, and that the work of Galileo and Descartes on the laws of motion were very much in his mind.

Of course Newton himself gave credit to Galileo for using the first two laws in describing falling bodies, credit that Galileo did not really deserve, while at the same time he studiously avoided mentioning his more important debt to Descartes. Even in the nineteenth century a Newton-worshiper like William Whewell could write:

> The *Principia* can hardly be said to contain any new inductive discovery respecting the principles of mechanics; for though Newton's *Axioms* or *Laws of motion*, which stand at the beginning of his book are a much clearer and more general statement of the grounds of Mechanics than had yet appeared, they do not involve any doctrines which had not yet been previously stated or taken for granted by other mathematicians.[2]

What, then, was "new" in Newtonian mechanics? Professor Westfall argues, and I believe correctly, that Newton's greatest contribution was an improved understanding of that terribly slippery concept, the notion of a force. It was a concept that Newton understood much better than Whewell, which helps to explain why Whewell underrated his illustrious fellow countryman in the passage just quoted.

In his paper Professor Westfall says that Newton was "the first man fully to comprehend . . . that the prime necessity of an operative dynamics was a conceptual unit to measure the 'external cause' of changes of motion"[3]—that is, a force which was not the Galilean impetus or the Cartesian "force of a body in motion," but an external force producing a *change* in momentum.

Newton's dynamics led him steadily in this direction, but his antipathy to Descartes and his insistence on absolute motion prevented him from discarding a force inherent to matter. Thus in spite of the logical direction of his dynamics, Newton retained two forces even in the *Principia*, one external force impressed on bodies to change their motion, and another internal force of inertia that resisted any change in the body's state of rest or uniform motion. In the later manuscripts and in the *Principia*, this inertial force of inertia no longer served to distinguish between relative and absolute motion, and appeared as a force only when the body's motion was altered by some external cause. The paradoxical nature of this latter force is aptly illustrated by Professor Westfall's phrase "the ertness of inertness." There is indeed something contradictory in the notion of a power of passivity.

I hope that I have correctly stated Professor Westfall's conclusions. These are difficult ideas. Newton wrote very clearly on the laws of motion, but generations of historians have been able to misunderstand him; and I am sure that I am equally capable with respect to Professor Westfall's paper. But assuming that my interpretation is somewhere near accurate, let me attempt a few comments.

One of the most revealing aspects of Newton's early mechanical manuscripts is his emphasis on the dynamic nature of force. Force was the cause of a change of motion; in the *De gravitatione* it also named an internal power that continued a body's state of rest or uniform motion. In either case it was a real entity quite apart from the motion or change of motion that it produced.

The later reasoning, which began with d'Alembert and Maupertuis, and which declared that force was merely a name for an observed change of motion, was foreign to Newton's philosophy. This is all the more dramatic when one realizes that in the treatment of celestial mechanics he could have dispensed with forces altogether. The law of gravitation could be stated in terms of accelerations and the only dynamic notion required was that of mass. As long as Newton did not have to deal with cases of static equilibrium, the concept of force was not really necessary. But the possibility of eliminating forces obviously did not enter Newton's mind. Change of motion, for him, required a real physical cause, and that cause was the force.

That is why I believe that Newton would have been unhappy with Professor Westfall's frequent statement that force in Newton's writings was to be *measured* or *defined* by change of motion or rate of change of motion.[4] Newton certainly wished to make force a quantitative concept, but I doubt that he wished to have it measured by its effects. If this were the only way that force could be measured, then it would be the only way in which it could be defined quantitatively.

Of course Newton frequently stated that impressed forces were *proportional* to the changes in motion that they produce, but this is different from saying that forces are *measured* by those changes. There are statements by Newton that might be interpreted to mean that changes of motion measure the forces (I have in mind his definitions of absolute, motive, and accelerative centripetal forces in the *Principia*),[5] but it seems to me that even in these cases Newton intends to describe a *proportion* between a real cause and the motion it produces rather than a definition or measure of the cause in terms of the motion.

This leads directly to Newton's insistence on the *absolute* nature of motion. The *De gravitatione* is largely taken up with a philosophical criticism of Descartes' concept of relative space and relative motion, but it also contains Newton's argument that a real change of motion is always accompanied by a force. The most obvious example is circular motion, but in the *Drafts of Definitions and the Laws of Motion* and more obviously in the *Principia* he adds that any change of motion—acceleration in a straight line for instance—is accompanied by a measurable force.[6]

Now this force that distinguishes real absolute change of motion from apparent change of motion is obviously not to be measured by the change of motion. In that case it would not distinguish between a real change and an apparent change. Instead it has to be measured by the increased tension in a string as in the case of Newton's rotating globes, the deformation of colliding spheres, or the sudden apparent freedom from gravitation of a body accelerated in free fall. It seems to me that this is a far more general phenomenon than the "narrowly confined set of absolute rotations" to which Professor Westfall refers.[7]

From his early writings the reality of forces and the essentially dynamic nature of Newton's mechanics convinced him that there is a difference between a real change of motion and only an apparent change. Unfortunately the same argument does not apply to bodies in uniform motion. No forces tell us which objects in uniform motion are really moving and which are at rest, and this is certainly one reason why Newton could not easily abandon the *vis insita*.

In the *Principia*, however, Newton argues for the absolute nature of uniform motion not from the existence of a *vis insita*, but from his belief in the reality of absolute space. By the time he wrote the *Principia* the *vis insita* could no longer distinguish between absolute and relative motions. Therefore in his mature mechanics, the mechanics of the *Principia*, the external impressed force was a more convincing argument for absolute motion than the *vis insita*. Newton must have realized that he did not have the complete answer, and it was probably wishful thinking that led him to state at the end of the definitions in the *Principia* that the method of obtaining true and absolute motions "shall be explained more at large in the following treatise. For to this end it was that I composed it."[8]

The insistence on absolute motion may be interpreted as an act of defiance, as Professor Westfall has done, or as a result of Newton's theological convictions. The long *Scholium* on absolute space in the *Principia* seems to me, however, to be the statement of a scientist, who feels that he has the proof of a cherished idea almost within his grasp. Motion in *general* was certainly not relative; he could demonstrate this fact for circular and accelerated motions and for the motions produced by impact. Would it be possible to prove the absolute nature of *all* motion from dynamics? The simplicity and uniformity of nature would seem to require it.

The failure of the *vis insita* to clinch his case was a disappointment. The old inherent force which appeared in the *De gravitatione* and the first version of the *De motu* would have successfully done the job if it had a measure independent of the motion it produced. Since it had no such measure it could best be sacrificed. But Newton remained optimistic. He had found a real dynamics of forces far superior to Descartes' mere translations, and he could expect that the wonderful harmony of nature would provide that small final piece of evidence that he needed to demonstrate the universal reality of absolute motion.

As for the problem of deciding whether force should be taken proportional to change of motion or rate of change of motion, I agree that Newton never answered this question and I also agree that it was not a great handicap in the practical application of his mechanics.

But it remained as a dilemma for physicists of the eighteenth century as it had been a dilemma for Descartes and Galileo, and I cannot believe that it was not also a dilemma for Newton. Certainly he must have been aware of the old problem of the apparent incommensurability between impacts and pressures. Maybe he *intentionally* left the concept of force vague enough to cover the three kinds of impressed force that he mentioned in the *Principia*. Maybe he finally realized that in some scientific quarrels, such as the existence of hard bodies and the *vis viva* controversy, when you don't have the answer, it is possible, maybe even better, to keep quiet.

1. Richard S. Westfall, "The Problem of Force in Galileo's Physics," *Galileo Reappraised*, ed. Carlo L. Golino (Berkeley, University of California Press, 1966), 67.

2. William Whewell, *History of the Inductive Sciences from the Earliest to the Present Time*, 3rd ed., 3 vols. (London, John W. Parker and Sons, 1857), I, 355.

3. Richard S. Westfall, "Stages in the Development of Newton's Dynamics," presented at the *University of Oklahoma Symposium in the History of Science and Technology*, Norman, Oklahoma, April 8–12, 1969.

4. *Ibid.*, 5, 7, 8, 10, 14, 17.

5. Isaac Newton, *Mathematical Principles of Natural Philosophy and his System of the World*, ed. Florian Cajori (Berkeley, University of California Press, 1960), 4–5.

6. John Herivel, *The Background to Newton's* Principia. *A Study of Newton's Dynamical Researches in the Years 1664–84* (Oxford, Clarendon Press, 1965), 310; and *Principia*, 10–12.

7. Westfall, "Stages . . . ," 22.

8. *Principia*, 12.

Commentary on the Paper of Richard S. Westfall

By Gerald C. Lawrence, University of North Dakota

I THINK that a reasonably accurate statement of Professor Westfall's thesis would be that Newton's work on the revisions of the *De motu* during the period 1684–85 was essential to the creation of the *Principia*, in that, prior to this time, the conceptual apparatus at Newton's disposal was inadequate for the construction of a workable dynamical theory, and in that, in the course of revising the *De motu*, a suitable conceptual apparatus was evolved. Further, the key relationship of the new conceptual apparatus was that of equilibrium between impressed and inherent, external and internal force.

In the course of developing his thesis, Professor Westfall has shown several major conceptual difficulties that had to be overcome by Newton, all of which center around the idea of force.

First of all, Newton inherited the notion of force inherent in a moving body and responsible for its motion, a force, which is thus measurable by the quantity of motion, *mv*.

This idea of force contrasted with the idea of force implied by the principle of inertia. Clearly, if bodies maintain a given motion unless affected by an external cause, then force, as that cause, must produce a *change* of motion, a Δ*mv*.

Finally, a third idea of force is that related to the changes of motion involved in circular motion and free fall, i.e., uniformly accelerated motion, where force, as cause, is measured by *ma*.

The attempt to deal with these various ideas of force, which are clearly dimensionally disparate quantities, involved Newton in the question as to the absolute or relative character of motion, and hence of space and time, since, only if motion were absolute, could force as a cause of motion possibly be real, and thus related to force as a cause of change of motion.

However, absolute motion could only be distinguished from relative motion through force, and this force itself had to be the essential, ultimate reality of the world. This led to a new conflict because Newton, unlike Leibniz, was not willing to relinquish the idea of passive, impenetrable, hard particles of matter, and thus a problem arose over the essential activity or inactivity of matter. This, as Professor Westfall pointed out, was to be neatly handled by Newton through the idea of the "ertness of inertness."

All of these problems confronted Newton when he began work on the *De motu*, and thus the solution, or possibly circumvention, of them in the course of the revisions of the *De motu* constituted a vast step toward the creation of the consistent dynamical theory of the *Principia*.

According to Professor Westfall the key to this development lies in the principle of inertia. It was to be through the transformation of the concept of inherent force according to the dictates of the inertia principle that the confusion over forces could be resolved into a relationship between "impressed" force and inherent force.

In the revised *De motu*, the relationship between the forces found expression in the hypothesis of the force parallelogram, but even here the dimensional disparity seems not to have been overcome. Newton combined *mv* forces with Δmv forces in the same parallelogram. Nonetheless, by the third revision, the force relationship had crystallized into a statement about change of motion and impressed force that is virtually identical with the Second Law of the *Principia*. The Second Law, however, implies the concept of inertia, which contradicted the still retained idea of inherent force as responsible for uniform rectilinear motion, and its corollaries of absolute motion and space and of an essential activity of matter.

Ensuing revisions of the *De motu* zeroed in on the notion of inherent force; Newton stated its proportionality to the quantity of motion, and redefined its action. Instead of maintaining uniform rectilinear motion, in its new character, it resisted efforts to change the state of motion, a change of conception clearly in accord with the inertia principle.

Subsequent revision elevated inherent force to a universal property of matter, and equated it with the inactivity of matter, *vis inertiae*. This new property of matter could now serve as the logical link between change of motion and impressed force, defined as that by which a body is urged to change its state of motion.

These two forces, impressed force and inertial force, and their constant relationship of equilibrium then served as the core of a new dynamical theory. Elaborated into the famous Laws of Motion, they constituted the basis for the powerful theoretical structure presented in the *Principia*.

However, in spite of the important modifications in the concepts of inherent and impressed force and their relationship through the inertia principle, a number of problems remained unsolved. First of all, the central role assumed by the inertia principle removed the necessity for maintaining the absolute character of motion and space. And yet, as Professor Westfall has stated, "In the process by which *De motu* expanded into the *Principia*, the space devoted to the argument for absolute motion grew in size with every change in Newton's dynamics that diminished its operational relevance. Newton's assertion of absolute motion has all the appearance of an act of defiance hurled in the face of the very current of thought on which his dynamics itself was borne inexorably toward its ultimate form."

A second problem has to do with the dimensionality of force. The second law asserts impressed force is proportional to change in motion, Δmv, and not ma. Professor Westfall has stated, "I am convinced that the dilemma has no resolution in Newton's works, that the problem visible in the *Waste Book* remained a problem unsolved, and that he used the term force indiscriminately both in contexts where it meant Δmv and in others where it meant ma." In other words, Newton apparently made no distinction between "force" and "impressed force," and none was essential to his dynamics, the key distinction in dynamics being that between inherent and impressed force, between external and internal force.

Thus, while Professor Westfall has demonstrated his contention that theoretical developments of primary significance to Newton's mature theory of mechanics took place in the years immediately preceding the writing of the *Principia*, he has left two major inconsistencies unexplained except on irrational grounds; an act of defiance in the case of absolute motion, and indiscriminate usage in the case of the term "force."

In view of the precise, logical, and altogether imposing character of the *Principia*, it seems to me that the persistence of these inconsistencies in the analysis of the development of Newton's thought indicates that possibly something has been omitted from the analysis. There are two additional considerations that should be brought to bear on the development of Newton's dynamics which can do much to integrate its apparent inconsistencies into a tight logical structure. First one must inquire as to the abstract relational structure, or the abstract calculus, of the theory, and second one must inquire as to the empirical significance of theoretical terms of the theory.

As to the abstract relational structure of Newton's dynamics, I would like to put forward the idea that it is identical with Newton's own calculus of fluxions and that a distinction between force as measured by Δmv and force as measured by ma was made by Newton in line with relationships characteristic of the calculus of fluxions. To be exact, the two are related in the same fashion as a "fluent quantity" to its "fluxion."

To find some support to this assertion, it is necessary to look at the relation between force and motion in the *Principia*. The approach to the concept of force and to its relationship with motion can be made through the postulates of the theory and the accompanying definitions in the *Principia*. For my purpose here it will be sufficient to look at the Second Law, which states that the change of motion is equal to the motive force impressed.

The definition of impressed force is of little help in understanding the law. "An impressed force is an action exerted upon a body in order to change its state, either of rest, or of motion in a right line." All that emerges from this definition is that impressed force is an action of some agent as well as being proportional to a change of motion. The agent producing impressed force and change of motion must be the force which is the cornerstone of the theory. To get at that agent, however, it is necessary to have an independent determination of impressed force, one that does not involve change of motion.

The definition of the motive quantity of a centripetal force provides the necessary separate determination of impressed force. It is the measure of centripetal force and is proportional to the motion which it generates in a given time. Newton immediately identified this quantity as weight, so that motion, and hence impressed force, are generated by weight, or some weight-like endeavor in time.

Leaving aside any further pursuit of the idea of weight for the moment, the Second Law now has the meaning that a weight-like force, multiplied by a time during which it acts produces a change of motion. But this could be strictly true only for a constant force. If force varies through space, then the Second Law in the above form could be true only for very short intervals of time, which suggests that the Second Law is basically a relationship between indefinitely small or infinitesimal quantities.

The infinitesimal character of the theory can be further demonstrated through the first two corollaries to the Laws of Motion. Corollary I concerns the well-known parallelogram of forces and states that a body acted upon by two forces simultaneously, will describe the diagonal of a parallelogram in the same time as it would describe the sides by those forces separately. The proof of the corollary is carried out with forces impressed in a point and with the constant velocities associated with them.

The significance of Corollary I is made clear in Corollary II, which begins "and hence is explained the composition of any one direct force . . ." Corollary I is the proof of Corollary II. However the direct forces of the second corollary are weights, so that the question arises as to the justification for the extension of the results of the first corollary to the second.

In the first, constant velocities are conceived as lines generated by points moving with constant velocity. These lines determine a parallelogram, and hence a diagonal which expresses their relationship. This relationship holds good for the motions thus represented and also for the impressed forces proportional to them. However, impressed force has been seen as the product of weight and duration; conversely, weight is the instantaneous increment of impressed force. Therefore, by extending Corollary I to application in the situation of Corollary II, Newton proclaims that Corollary I is concerned with an instantaneous event.

Thus, while the basic relationship of Newton's dynamics, the proportionality between the impressed force and change of motion is indeed based on the equilibrium between impressed and inertial force, it also represents an instantaneous event. For this reason, weight as the instantaneous increase of impressed force, or its fluxion, can be equated with ma, the instantaneous increase, or fluxion, of the motion, and this relationship, involving force as weight is separate from the Second Law, involving impressed force, but is logically related to it through the calculus of fluxions.

In order to make use of the relation between weight and ma in the treatment of the planets, Newton substituted, through the gravitational hypothesis, a space function for centripetal force. What justification did he have for so doing? If one turns to the definitions relating to centripetal force, one finds that the weight of a body is the product of its quantity of matter and an accelerative force characterizing the *space* occupied by the body. Space itself thus appears as an agent, and active principle, a substance. This being the case it must be of an absolute character. The universal inactivity of matter dictates that space, by default, must be the active principle.

Thus, in spite of the fact that the transformation of the inherent force of a body into inertial force eliminated one reason for an absolute space, the need for a separate and empirically significant determination of centripetal force, in combination with the inertia of matter, dictated another.

By way of conclusion, I would like to offer a modification of Professor Westfall's conception of Newton's struggle to produce a workable theory of dynamics. I think that Newton's calculus was, from its inception, intended to be a means of representing mathematically the real causal structure of the universe, that is, the motions of indefinitely small particles as the ultimate cause of all phenomena. Consequently, Newton's task in the construction of a workable dynamics was to fit dynamical concepts into the logical, relational structure represented by the calculus.

The key to his final theoretical achievement was indeed the idea of equilibrium between impressed and inherent force conceived in accordance with the principle of inertia. Only this enabled Newton to set up a proportionality between impressed force and change of motion, and hence between force and *ma*, for as Newton wrote in his *Treatise on the Method of Fluxions*, "things only of the same kind can be compared together, and also their velocities of increase and decrease."

Uniformity and Progression:
Reflections on the Structure of Geological Theory in the Age of Lyell

By Martin J. S. Rudwick, University of Cambridge

"I ALWAYS FEEL," wrote Darwin in 1844, "as if my books came half out of Lyell's brain." No comment summarizes more succinctly the immense influence of Lyell's writing on the scientific thought of his age. Few men of science working in natural history—a term not yet become derogatory—were unaffected by the persuasive impact, imaginative power and sheer scope of Lyell's work. But the very magnitude of Lyell's achievement raises acute problems of interpretation.

How revolutionary was Lyell's geology? Did the *Principles of Geology* reform that science—or even create it as a scientific discipline—in the manner that Lyell intended? Was its argument persuasive to all whose minds had not been made impervious by earlier modes of thought? How much did Lyell's approach to geology differ from that of his contemporaries? In asking such questions I am not trying to play the fruitless game of devaluing a great man's achievements by pointing to his "forerunners." I want instead to ask what I think is a legitimate and important historical question: what is the contemporary scientific context of Lyell's geology?

I will start by quoting an opinion which deserves respect, for it comes from a scholar who is making a comprehensive study of Lyell.

> The year 1830 [writes Professor Leonard Wilson] marked a great divide in the history of geology. Before that year there had been many great accomplishments, but there had been no critical assessment of the meaning of geological phenomena . . . the interpretation of the past history of the earth . . . remained fanciful and speculative. After 1830 geology became a science.

Here we have a clear statement of a radical claim on behalf of the *Principles of Geology*: namely, that the year of publication of the first volume marks a genuine historical watershed. To use the cur-

209

rently fashionable jargon, it is a claim that Lyell's work did not merely set up a new paradigm for research—as Professor Kuhn himself has suggested—but also that it formed the *first* paradigm in geology, and that it was preceded only by work which was in some sense not "scientific." This claim, if valid, would be historically of the greatest importance, and therefore merits careful consideration.

Certainly there is much in Lyell's own writing to suggest such an interpretation. Lyell frequently claimed that any departure from his conception of the uniformity of nature was "unphilosophical," or at least that it had an effect that was the reverse of heuristic. But it was not for nothing that Lyell's early training was as a lawyer, or that his contemporaries saw in the *Principles* all too much of the "language of an advocate." Lyell's polarization between himself and his opponents—a polarization he took so far as to express it in the Pauline imagery of a cosmic battle between the forces of good and evil— certainly tells us much of importance about the complex character of Lyell himself. But there is a danger here that as historians we may allow ourselves to be swayed—perhaps even brainwashed—by Lyell's persuasive oratory. We need surely to try to get behind the simplicities of the courtroom drama to the more complex, less clear-cut situation that generally exists before the case comes to court.

Unfortunately we are not aided in this task by the categories we have inherited from the period. That Victorian polymath William Whewell can hardly be blamed for the fate of the scientific terms he coined. But it is increasingly clear that the labels with which he characterized the Lyellian debate have now become a pair of polysyllabic millstones around the neck of the history of geology.

"Uniformitarianism" and "Catastrophism" were useful terms to catch some of the striking features of the debate; but they left unclear the meaning of "uniformity" and the importance of "catastrophe," and more seriously they accepted too uncritically the battle-lines that Lyell had drawn. Moreover, since Whewell's time there has been a tendency for the terms to degenerate into mere labels of—respectively —praise and abuse. In our present age of enlightened historiography, historians of science may be unwilling to admit openly to indulging in such "Whiggish" activity; but I think it is clear that if we are to get out of this stalemate in the history of geology we must first agree to abandon the use of Whewell's terms, except within the strict meaning and historical context in which he first applied them.

That negative prescription must surely be coupled with a positive one: we need to make a semantic analysis—however simple-minded it may be in the eyes of philosophers of science—of the concept of "uniformity" in geology, and of the concepts to which it may stand opposed. No one who attempts such an analysis can do so without feeling a great debt to the pioneer work of Professor Hooykaas in this field; and my own indebtedness, in what follows, will be clear enough to all who are familiar with his work. Following the sincerest form of flattery, however, I want to suggest that his analysis can be taken further, and that this will help to clarify the historical questions I started with.

The problem of uniformity arises in geology from its membership of what Whewell usefully classified as "palaetiological" sciences: it is concerned with reasoning about the nature and causation of events in the past. That Lyell clearly recognized the implications of this classification is shown by his frequent use of illustrative analogies drawn from the historical and archaeological studies of his own day. He also recognized that in geology, as in other palaetiological studies, there must be a basic commitment to a belief in the uniformity of the fundamental "laws of nature" (whatever those are!), since without this recognition all research would be impossible. But for his own polemical purposes he often chose to imply that just this basic commitment was lacking in his opponents; whereas in fact they adhered as strongly as he did to "uniformity" in this sense.

The debate only becomes philosophically interesting at a higher and more complex level of causation, that is, in considering the uniformity of geological processes or agencies. Such "causes," as they were termed—for example marine erosion or vulcanism—were agreed to be conformable to, and indeed based upon, the physico-chemical "laws of nature"; but they were evidently causes of far greater complexity. Only about these geological causes can a semantic analysis usefully be made.

I think we can identify *four* logically distinct types of uniformity on this level. All four, and their opposites, were discussed in the Lyellian debate, but Lyell himself either confused them, or believed that each was logically entailed by the others, or—most probably—he was convinced that the reforms he was advocating could only be secured in practice by holding firmly to "uniformity" all along the line.

First, there is the *theological status* of a past geological "cause," in relation to the creative activity of God. It might be *naturalistic*, achieved by "secondary" or "intermediate" means and therefore potentially within the realm of positive knowledge, even if the nature of the means remained obscure. This was not of course incompatible with a belief that its action was divinely sustained and providentially ordered. Or it might be *supranaturalistic*, not attributable to any secondary means, and therefore referable to a "primary" act of "creative" power.

Second, there is the *methodological status* of a past geological "cause." It might be *actualistic*, corresponding in some way to "actual" "causes now in operation," so that observation of those causes could provide a reliable guide to its interpretation. Or it might be *non-actualistic*, in that it failed to have any valid analogy with the present day, being either a "former" or "ancient" cause that had ceased to act, or an intermittent cause that happened not to have operated during the short span of human history. Then interpretation in terms of actual causes would simply be misleading.

But it is important to note that actualism and its opposite can both be applied with varying degrees of vigour: a cause might be unobservable at the present day in the *degree* to which it formerly acted, and yet be perfectly actualistic in the sense that in *kind* it continues to act, perhaps on a much smaller scale. Obviously there is no sharp boundary in practice between degree and kind, since it is merely a matter of definition at what point a past cause becomes so different in degree from its present analogue that it deserves to be called different in kind.

Thirdly, there is the *rate* at which a past geological "cause" may have acted. It might be *gradualistic*, being perhaps insensibly slow within a man's lifetime, or even within the span of recorded human history. Or it might be *saltatory*, or in more familiar terms "paroxysmal" or "catastrophic," acting relatively suddenly and on a large scale. Here again there is clearly no dividing line between these terms, which are merely end-members of a continuum of varying rates or intensities.

Fourth and last, there is the overall *pattern* of a past geological "cause," when its action is traced over the whole known time-span of earth-history. It might be *steady-state*, exhibiting relatively minor fluctuations around a constant mean, at least when the overall state of the whole globe is taken into account. This is clearly the original meaning of "uniformitarian." Or the pattern might be *directional*, or in more familiar terms "developmental" or "progressive," in that, underlying the local fluctuations in its activity, a general overall trend can be detected on a global scale.

Here there are four meanings of geological "uniformity," and their opposites. It is important to see that they *are* logically distinct. A coherent geological synthesis could be constructed from *any* combination of these four pairs of categories. Moreover, there would be no logical absurdity, or empirical improbability, in assigning different causes to different categories in this scheme.

Using this analysis, I want to suggest for discussion that Lyell's *Principles of Geology* was not launched upon a pre-paradigm world of random observations whose meaning was not clearly understood, or upon a geology that had no articulated methods. I want to suggest instead that the scientific context of Lyell's work is *another* geological synthesis, of equally great scope, sophistication and explanatory power; a synthesis that shared the essentials of Lyell's method but differed significantly in its conclusions.

If this is true, why has this alternative paradigm not been more fully recognized? I think there are several reasons for its neglect by historians.

First, as I have already suggested, Lyell's eloquent but distorted account of his predecessors is highly persuasive. His readers are all too readily swayed into believing that Lyell alone, apart from a few commendable forerunners, was bringing light into a dark world of obscurantism and prejudice.

Second, Lyell's account has some slight plausibility if historical attention is confined to geological work in the English language. But geology was self-consciously an international activity, and England was not the center of the scientific world in the 1820's.

But third, the synthesis that Lyell's work opposed lacks the most desirable element for historical identification: a great name. It was a coherent research tradition, but it is expressed more adequately in the scores of papers in the French, German, and English scientific periodicals, than in any single book or in the work of any single geologist.

If this paradigm lacks a great personal name to identify it, what are we to call it? To term it "catastrophist" would be to give too great prominence to its saltatory elements, besides introducing extraneous overtones of calamity and disaster. To term it "progressionist" is equally misleading, because the overtones of improvement in that term are only applicable to certain aspects of the theory. I shall use instead the more neutral term *directionalist*. This will emphasize at once what I believe to be the most significant element in it, namely the directional pattern that it traced within the whole time span of earth-history.

The foundation of this directionalism was an essentially geophysical theory. Buffon's *Epoches de la Nature* had long before popularized the idea of a gradually cooling earth. But such theories gained great scientific prestige in the 1820's from two related directions.

First there was the increasing empirical evidence for the reality of a "central heat" in the earth. It had long been known that the temperature in mines increased steadily with depth, but there were many objections to the interpretation of this as evidence for a central heat. In the 1820's, however, Louis Cordier's rigorous analysis of the evidence proved clearly that the geothermal gradient was not only real but universal. This suggested very strongly that its source must be very deep-seated and global in extent, although local variations in the gradient were fairly clearly related to volcanic activity.

This empirical work seemed to establish the reality of the central heat; but its interpretation as a *residual* heat was powerfully supported by the scientific authority of Fourier's physics. Applying his heat theory to the problem of a cooling earth, Fourier showed that it could account for the observed geothermal gradient. At the same time it could explain why there had been no perceptible change in the earth's diurnal rotation, and therefore in the earth's radius, since the time of the Ancients. At earlier geological epochs the rate of cooling of the earth, and therefore the rate of contraction, would have been relatively rapid; but these rates would have diminished progressively. Thus the temperature at the earth's surface, plotted against time, would have fallen along an exponential curve; and in more recent epochs the heat-loss from the earth's interior would have been very small compared to the effects of solar radiation, so that conditions would have become fairly stable and uniform.

It is only in the light of this theory of a residual central heat that we can understand the directionalist interpretation given to other phenomena.

For example, the earlier controversies on basalt had been settled effectively in favor of its volcanic origin. But this only served to emphasize the wide extent and extremely large scale of volcanic activity at earlier epochs: there was no modern vulcanism to match the scale of the great basalt flows of Iceland and the northwest of the British Isles.

This alone would have suggested that vulcanism might have diminished in the course of time; but such a conclusion was powerfully strengthened by the belief that only the earth's central heat offered an adequate source for volcanic heat. If the central heat was residual, what was more natural than to conclude that its surface manifestation in vulcanism would have decreased correspondingly? This line of reasoning underlies all the directionalist volcanic theories of the period, including for example those of Lyell's friend George Poulett Scrope and Buckland's colleague Charles Daubeny.

The theory of a cooling earth implied that conditions on the surface and in the earth's crust would have been very different in the earliest epochs from what they later became. It should be unnecessary to point out that this did not imply any suspension of the laws of physics and chemistry. It did imply, however, that in remote epochs some processes might have occurred on a large scale, which now could only be imitated with difficulty—if at all—in the laboratory.

This was particularly relevant to the problem of the origin of granite and gneiss. Granite, for example, was known to be sometimes intrusive into earlier rocks, as Hutton had showed long before. But such occurrences were notably commoner among the oldest rocks, and became progressively rarer among the more recent. Since it was agreed that granite was the product of hot conditions in the depths of the earth, it seemed natural to conclude that its formation was a process that had acted on a large scale in the earliest epochs, and had become progressively more restricted as time went on.

But not all granites could be interpreted so simply: many seemed to grade insensibly into banded gneisses, and thence into schists and even more clearly sedimentary rocks. These complex masses of unfossiliferous rocks generally underlay even the Transition strata, so it was reasonable to attribute them to conditions peculiar to very ancient periods of the earth's history. Under the hot conditions then prevalent, it was argued, such rocks could well have been formed by crystallization from a melt, or by precipitation from hot solutions. With further cooling, these rocks of chemical origin would have been replaced gradually by more normal sedimentary rocks of mechanical origin. In such rocks of the Transition series the earliest traces of organisms were found.

The mention of fossils brings us to one of the most important features of the directionalist synthesis. It is well known that a "progressionist" history of life was one of Lyell's main targets for attack. But this biological interpretation was extremely closely integrated with the view of the history of the earth which I have just outlined.

Directionalism in the organic world did not merely run parallel to directionalism in the inorganic world; there was a kind of causal relation between them. In effect, the Cuvierian view of the close integration of the organism to its environment was applied to the concept of a directionally changing environment, and so became a means of explaining the directionally changing nature of successive faunas.

The gradually cooling globe had presented a succession of different environments, at first too hot for life of any kind, then (as we would say) hypertropical, then tropical in all parts of the earth, and finally approximating to its present condition. In this succession of environments, appropriately adapted organisms had successively come into existence, had flourished for a time, and then, as the environment continued to change, they had become extinct and had been replaced by others better adapted to the new conditions. Moreover, it was commonly believed that this succession of environments had been such as to allow progressively higher and more elaborate forms of life to exist.

In this way, a directionally changing environment could explain the empirical phenomena of characteristic fossils, on which the spectacularly successful development of stratigraphy was increasingly based. But it could also explain the phenomena of the progressive development of life.

It was not surprising, on this view, that Man had appeared very recently on Earth, that mammals were confined to the Tertiary strata, that reptiles extended much further back into the Secondary strata, or that the Transition strata seemed to contain few if any vertebrates at all. This view of the adaptation of the forms of life to directionally changing conditions was so firmly held that there was much skepticism when an anomaly was reported.

Some small mammalian jaws had been found in the English Secondary strata. The discovery was authenticated both by the detailed stratigraphy of William Smith's home area, and by the authority of Cuvier's vertebrate anatomy. Yet it was so unexpected that both the stratigraphy and the anatomy were challenged. This example is significant, for the resultant controversy was conducted entirely within the directionalist paradigm; and the anomaly was finally seen as an unexpected confirmation of the progressive development of life. The fossils were of *marsupial* type, and not placental mammals; so it did not seem surprising that this "lower" form of mammalian organization had appeared on earth before the true mammals of the Tertiary epoch.

Thus the increasing elaboration and diversity of animal life, when traced through the history of the earth, was seen as a correlate of the directionally changing environment. This view of the history of life was confirmed unexpectedly in the 1820's by Adolphe Brongniart's pioneer work on fossil plants; for this proved that the earth's floras, like its faunas, had shown increasing elaboration and diversity in the course of time. Moreover, Brongniart's work provided further confirmation of the theory of a cooling earth.

The early floras of the Coal period were analogous in general character to some of the tropical floras of the present day, yet their fossil remains were found even in cool-temperate latitudes. This seemed good evidence not only for a higher temperature at that period in the past, but also for a more uniform temperature all over the earth's surface. This was precisely as expected on Fourier's heat theory, for the present climatic zonation of the earth was seen as a consequence of the present regime of solar radiation, whereas at earlier epochs this would have been masked by the greater heat flow from the earth's interior.

A "progressive" succession of adapted forms of life was thus well-established in the 1820's. What remained obscure, of course, were the means by which these forms had appeared on earth and later disappeared. Extinctions could still be explained by invoking Cuvierian *révolutions* (to which I shall turn shortly), but the problem of origins could no longer be evaded simply by invoking the mechanism of faunal and floral migrations.

Generally, of course, naturalists were content—as Lyell was too— to leave the problem unexplained, and to use the deliberately non-committal language of species "coming into existence" at successive epochs. Yet it is important to notice that when this empiricist agreement not to speculate on the problem was defied, as it was by Étienne Geoffroy St. Hilaire, his mechanism of transmutation by the influence of the environment was fully integrated into the directionalist synthesis. Geoffroy had no need to invoke the primary Lamarckian "tendency to improvement," for unlike Lamarck he accepted the idea of a directionally changing environment. Geoffroy's transformism was geared to the evolution of the environment: successive forms of organization had developed in response to the successive changes in the environment.

So far I have emphasized the primarily directional nature of this synthesis of geology and biology in the 1820's. But it was as "catastrophism" that it later became known, so we must next consider the place of saltatory elements within it. How essential to the synthesis was the postulate of occasional sudden events of great magnitude?

I have already commented that the word "catastrophe" carries overtones of disaster which are largely irrelevant. But of course Whewell's choice of that word, when "paroxysm" and "revolution" were also available, does reflect the importance given to the most recent event of the kind, which had been widely believed to be the catastrophic Flood recorded in all the ancient literatures known at the time (not only the biblical). In fact, however, the identification of the geological Deluge with the Flood had generally been abandoned by the time that Whewell coined the term "catastrophism." It is probable, however, that he was conscious of other overtones in the word. He must have been well aware that in Greek drama the καταστροφή is the final dénouement which brings the action to its conclusion. Given that a dramatic physical event seemed to have occurred on earth in the geologically recent past, it would be natural for Whewell to view this as the καταστροφή of a long directional development of the earth and its inhabitants.

Certainly there *was* strong evidence for some highly peculiar episode in the geologically recent past. In assessing the diluvial theory of the 1820's, we must never forget that nature played a harsh trick on the geologists of the period. With the benefit of scientific hindsight we can see that there had indeed been an extremely unusual event. A glacial period is a rarity in the history of the earth; and since it came to an end (temporarily?!) a mere ten thousand years ago, its effects were bound to seem all the more dramatic.

The north German plain, for example, was strewn with boulders of distinctive rocks that could be traced right across the Baltic to their sources in Scandinavia. On a smaller but still impressive scale, such erratic blocks could also be traced across Britain, often perched on hill-tops and showing no relation to the present drainage system. Such phenomena clearly witnessed to a discontinuity between the present and the geologically recent past; they implied a period of conditions very different from those of the present day.

What agencies could have been responsible for these spectacular effects? It was no use appealing to the magnitude of geological time, for no amount of time could shift boulders the size of houses by the agency of the processes now at work in the areas concerned. Moreover, the organic world seemed to have been affected as radically as the inorganic: as Cuvier had shown, the "diluvial" gravels yielded a whole fauna of extinct mammalia. The only plausible explanation was that a violent rush of water had swept suddenly over the relevant land-areas, causing both the transport of erratic blocks and the annihilation of the indigenous fauna.

It is unfortunate that the provincialism of some English-speaking historians has made William Buckland seem to be the archetypal catastrophist. Buckland was a brilliant lecturer, an entertaining showman and publicist, but a somewhat mediocre geologist. His reconstruction of the diluvial cave fauna was generally admired; but his straining of the evidence, in order to prove the uniqueness, suddenness and universality of the diluvial episode, was equally generally criticized. During the 1820's, it gradually became clearer that not all the diluvium was of the same date, that several diluvial episodes must be postulated, and that none of them had been universal. But this did little to lessen the puzzle of the diluvium.

With such a dramatic event or events in the recent past, it was natural to infer that similar but earlier paroxysmal episodes might have been responsible for some other problematical phenomena. For example Leopold von Buch, one of the best geologists of the period, put forward his influential theory of *Erhebungscratere* (craters of elevation) to account for the puzzling features of volcanic calderas; only episodes of sudden elevation seemed adequate to explain them.

Similarly, when strata of sedimentary rocks were found in mountain regions torn and crumpled like putty, it was difficult to believe that any ordinary processes had been responsible, no matter how long the time that had been available. Such spectacular effects seemed to bear witness to occasional sudden episodes of mountain elevation. This was developed and given greater scope in one of the most important theories of the 1820's, Léonce Élie de Beaumont's theory of epochs of elevation. Élie de Beaumont used the latest refinements in stratigraphy to date the successive periods of orogeny (as we would say) with greater precision than ever before. This proved conclusively that mountain building had occurred at intervals throughout earth-history.

But Élie de Beaumont went further, and proposed a causal hypothesis for these successive paroxysmal episodes, integrating them into the directionalist synthesis. With a gradually cooling and shrinking earth, he believed that compressional strains would slowly build up in the earth's crust. When these strains exceeded the strength of the crust, it would suddenly shear and buckle along lines of weakness, and linear mountain chains would thus be elevated. The strains would thereby be relieved, but in due course they would build up once more, only to be relieved by another episode of mountain formation along some different set of directions. This hypothesis was extended —characteristically—to explain biological problems too. Sudden episodes of mountain elevation would inevitably cause huge tsunamis ("tidal waves") to sweep across the continents. This would naturally cause dramatic changes in the ecology of the animals and plants, and could be responsible for the apparently sudden episodes of mass extinction between successive series of strata. The diluvial episode became merely the last of many such occasional episodes, and was tentatively linked to the recent elevation of the Andes.

It was not by any quirk of jealousy or pique that Adam Sedgwick acclaimed Élie de Beaumont's theory while criticizing Lyell's: to Sedgwick as to most other good geologists of the time, Élie de Beaumont's theory seemed an exciting synthesis of hitherto unrelated problems, even if they had reservations about how far it would stand up to detailed testing.

In Élie de Beaumont's theory, then, we can see how occasional saltatory events could be generated from an essentially gradualistic background (i.e., the gradually cooling earth), without any suspension of the ordinary "laws of nature." In the intervals between these saltatory events, Élie de Beaumont believed there had been immensely long periods of tranquil conditions, during which the ordinary strata had been deposited. This should serve to emphasize how saltatory events were only invoked when and where they seemed necessary to explain the observed phenomena: no "catastrophist" believed the past history of the earth had been an unrelieved succession of violent events. To use the phrase of the time, they were no more "prodigal of violence" than the phenomena seemed to require.

Using the same epigram, it should be clear that they were not "parsimonious of time" either. Of course they did not use the vastness of geological time as a device for eliminating the suddenness of saltatory events, as Lyell was to do; but it by no means follows that they did not believe in the enormous time-scale of earth-history.

This is a difficult point to prove by quotation, because all geologists were justifiably reluctant to commit themselves to statements about the magnitude of geological time, when they had nothing but their geological intuition to go on. But I think it is clear, nevertheless, from frequent comments on the slow tranquil deposition of ordinary strata, that they envisaged an extremely lengthy time scale. Paroxysmal events were postulated, not in order to accommodate large effects within a constricted time-span, but because the nature of the effects themselves seemed to call for more than just the long-continued action of slow processes.

If there remains any residual suspicion that "catastrophism" was necessarily linked to a belief in sudden episodes of species-creation, that too should be dispelled by noting how completely Geoffroy St. Hilaire assimilated his naturalistic transformism to the "catastrophism" of his geological contemporaries. Having geared organic change to environmental change, he was able to use occasional episodes of rapid environmental change as an explanation for corresponding episodes of rapid organic change. The embarrassing difficulties of transforming one major type of organization into another, without unadapted intermediates, were thus at least alleviated.

One final point about the saltatory element of the directionalist synthesis. Although Élie de Beaumont and others believed that paroxysmal episodes had punctuated the whole of the earth's history, including the recent past, other directionalists stuck more closely to the implications of the theory of central heat, and believed that paroxysmal events must have become less and less frequent in the course of time, as the earth cooled towards its present virtually stable condition. Thus Poulett Scrope, for example, could be highly critical of some of the more sweeping assertions of the diluvialists, and yet feel no inconsistency about postulating paroxysmal events for the more remote past. It was only to be expected, in his view, that agencies such as volcanoes and earthquakes would have been active on a larger scale in the earlier periods of the earth's history, and that they would have tended directionally towards quiescence.

So far I have discussed the directional character of this synthesis of the 1820's, and I have pointed out that saltatory events were only postulated where they seemed necessary, against a background of generally gradualistic changes. We now have to consider whether even this limited use of "catastrophe" was, as Lyell alleged, radically unscientific because it failed to be actualistic in method. Did it leave the door wide open to unrestrained speculation, by failing to interpret the past in terms of the present?

I think it is very difficult to substantiate Lyell's allegation. Most of the research I have been describing was certainly done with actualistic intentions, and with a good deal more methodological sophistication than Lyell was prepared to allow. Even Cuvier had postulated sudden *révolutions* only because he believed that no actual causes could account for the observed phenomena. But Cuvier saw no inconsistency in using actualism to interpret other phenomena, such as sedimentation; and of course he also used an actualistic argument to refute Lamarck's transformism, pointing out that since transmutation had demonstrably *not* occurred during recorded human history, no amount of geological time could justify postulating it for the more remote past. So even for Cuvier, comparison with the present was the proper fundamental method for geology, as indeed for palaeontology too; and non-actualistic events were only to be postulated where the method of actualism failed.

Comparison with the present was, moreover, not merely good geological method: it was a natural corollary of the general directionalist synthesis. This may seem paradoxical unless we remember the implications of Fourier's view of an *exponential* fall in the surface temperature of the earth. All geological research tended to confirm this general picture of a gradual slowing down of the rate of change of conditions on earth. Consequently it was entirely reasonable that the span of human history should be, in general, a quite reliable key to the more recent epochs of earth-history.

But it did not follow that actual causes could be assumed *a priori* to be *totally* sufficient for the reconstruction of the past. For one thing, the further one penetrated into the past, the more conditions would have differed from the present, and therefore the more tenuous the analogy would become. Even if the *kinds* of processes had been the same, their *degree* of intensity might have been so much greater that their manifestations might have no close parallel at the present day.

For another thing—and this underlines again the immensity of geological time that was assumed—the period of human history was so short by geological standards that it could not be taken *a priori* as an altogether adequate sample of even the more recent past. For if there were in fact causes that only operated occasionally, such as epochs of elevation, they might never have come under human observation at all, and yet they might have been immensely important in shaping the earth's development.

So there was general agreement that actualistic comparison of past and present was the best policy for research, but that it could not be expected *a priori* to be totally adequate for geological explanation. It would be unhistorical to pick out a few figures in the 1820's as actualistic forerunners of Lyell, and to attribute a lack of scientific method to the rest. All geologists were actualists by policy; they varied only in the *extent* to which they believed the past could be interpreted by close analogy with present causes.

There was in fact a continuing lively debate on just this point. This is reflected in the Göttingen prize question which was won by Karl von Hoff, with his massive compilation of the historically authenticated effects of actual causes. Von Hoff emerges as a convinced believer in the power and efficacy of actual causes; but it should come as no surprise to find that he was nevertheless not convinced that *all* phenomena in the past could be explained in this way. His belief in the explanatory power of actual causes was perfectly consistent with his recognition that some more paroxysmal events might have occurred in the past. It was also of course consistent with his directionalism.

This can be seen equally well, for example, in Poulett Scrope's work. Scrope could argue for the heuristic value of actualism as strongly as Lyell was to do later; but this did not stop him advocating a directionalist volcanic theory according to which vulcanism had declined in intensity during earth-history. Moreover, when later he wanted to point to the sources of Lyell's actualism, he was content to couple his own work with that of Charles Daubeny, although Daubeny allowed an even greater role to paroxysmal events than Scrope himself wished to concede.

Finally, what of the theological status of this directionalist synthesis and of the causes it postulated? Were "catastrophes" unscientific, as Lyell again alleged, in that they opened the door to non-material causation?

Once again, I think this is difficult to substantiate. Even Buckland was prepared to suggest possible secondary causes for the diluvial episode, and the more typical "catastrophism" of von Buch and Élie de Beaumont was even more clearly naturalistic. If the saltatory events of earth-history were sometimes left with their causation unexplained, this is not because they were thought to be due to divine intervention, but simply because many geologists wisely concentrated on establishing that certain events *had* occurred, recognizing that their causation was a logically separate question that could best be left until there was better evidence to go on.

The same methodological point also explains in part the general reluctance to speculate on the origin of new species. Probably many naturalists *did* believe that the phenomena of adaptation placed this problem beyond reasonable hope of any explanation in secondary terms, but the example of Geoffroy shows once more that generalizations are hazardous on this point.

So with the important—but in a sense limited—exception of the origin of new species the geology and palaeontology of the 1820's were thoroughly naturalistic. But of course this did not prevent them being given a providentialist interpretation. In particular, the belief in a directional approximation to the present relatively stable state of things was readily assimilated to a belief that the earth had thus been gradually prepared until it was in a fit state to be the habitation of Man.

But while this may have been a powerful motive behind the work of some directionalists, it is possible that historical concentration on the English scene has led to an over-emphasis on the role of natural theology in the science of this period. The Enlightenment tradition in Continental Europe may perhaps have made natural theology a much less significant factor in the areas where most of the best science was being done.

To sum up this very tentative analysis of the synthesis into which Lyell's *Principles* was launched: it was a coherent synthesis of geology and biology, in which the earth and its inhabitants had undergone a series of *directional* changes, the rate of change generally decreasing as conditions approached those of the present day. Superimposed on generally *gradualistic* changes, occasional *saltatory* events were inferred where necessary; the last such event might have been geologically recent, but their frequency and intensity were usually assumed to have decreased in the course of time. *Actualistic* comparison with the present was agreed to be a heuristic policy for research, but could not be taken *a priori* to be a totally adequate means of explaining all events in the past, and particularly the more remote periods of earth-history. Finally, with the general exception of the origin of new species, all events were taken to be due to *natural* secondary causes, operating in complete conformity with the ordinary laws of nature, though their pattern might be interpreted in terms of providential design.

I have left myself with no time to attempt a corresponding analysis of Lyell's own synthesis. Briefly, however, we may note that although he was deeply committed to a thoroughgoing *naturalism*, he did not differ in this respect from other men of science. Even at the one point where many of them were inclined to fall back on non-material causation—namely the origin of new species—Lyell too was content to use the conventional language of "creation," leaving it to be inferred, as he later told Herschel, that he believed some unknown secondary cause to be responsible. It can be argued that Lyell could excuse his ambiguity on this point, because the exact mode of origin of species was peripheral to his task of establishing a more fundamental position, namely a steady-state terrestrial system.

Second, it is well known that a rigorous *actualism* is an important element in Lyell's synthesis. He insisted that past causes had been similar to present, not only in kind but also in degree. But this is only an extreme form of a methodological policy which was accepted in varying degrees by all his contemporaries. And in practice it seems to have been governed less by considerations of methodological rigor than by a concern to demonstrate that geological processes were not declining in intensity.

Thus, for example, causes that had never been observed in human history were still acceptable to him if they might possibly occur in the future. Likewise he was not averse to inverting his actualism on occasion, as when he postulated species-production as a rare but unobserved event, on the basis of the evidence that new species *had* been introduced in the past.

Third, Lyell is generally supposed to have been fundamentally committed to gradualistic explanation in opposition to any form of catastrophism. Yet even here, his position on the gradual/saltatory continuum was in fact some way from extreme gradualism, and was governed more by the canons of actualism and hence by the requirements of a steady-state system. Thus he initially rejected the alleged rise in the Baltic shorelines, in spite of its extreme gradualism, because he doubted its actualistic authenticity and because it failed to fit his own (more saltatory!) theory of elevation by earthquakes. Similarly he rejected the extreme gradualism of Lamarckian transmutation in favor of an unspecified process which was, by implication, considerably more saltatory in character.

Fourth and last, Lyell's fundamental commitment is, I believe, to be found in the steady-state system he propounded. It is only at this level that the conflict between his synthesis and that of the directionalists bears the marks of a genuine confrontation of irreconcilable paradigms. Only here do we find the characteristic rival interpretations of the same phenomena, the viewing of the evidence through alternative spectacles.

I think this is reflected in the way in which the *Principles of Geology* was received by the scientific community: its attempt to extend the heuristic value of actual causes was generally acclaimed, but its use of actualism in the service of a steady-state or uniformitarian system was equally widely criticized. For the entire structure of the *Principles* is governed by the overriding strategy of advocating such a system, and this seemed to Lyell's contemporaries to involve the most serious straining of the evidence.

Throughout the 1830's, in fact, while Lyell saw his actualistic policy used increasingly in geological research, his steady-state system gradually seemed less and less tenable in the light of that research.

Edouard Lartet's discovery of the first fossil primates within the Tertiary epoch, Louis Agassiz's delineation of an age of fish preceding the age of reptiles, Roderick Murchison's description of a vast Silurian system largely pre-dating even the age of fish—these and many other discoveries seemed to confirm still more clearly the essential validity of the directionalist view of the history of life, rather than Lyell's steady-state view. Lyell's steady-state system for the earth itself fared little better, because he was unable to point to any indefinitely renewable source for the earth's internal heat, on which his system of balanced forces ultimately depended.

In conclusion, therefore, I suggest that however Lyell's work is evaluated—and I have scarcely touched on this question—it must be seen in the context of a rival synthesis of great scientific power and sophistication. Directionalism gave meaning and coherence to a vast range of empirical observations, and it was based on the best physics and chemistry of the time. Its theories were neither fanciful nor speculative: on the contrary they were consciously kept within the strict empiricist limits of "positive science." Lyell may have wished, as he said, to "establish the principle of reasoning in the science," but to his contemporaries the principles were already secure.

References

Brongniart, Adolphe, Nov. 1828. "Considérations générales sur la nature de la végétation qui couvrait la surface de la terre aux diverses époques de formation de son écorce," *Ann. sci. nat.*, *15*, 225–58.
———, 1828–[37]. *Histoire des végétaux fossiles, ou recherches botaniques et géologiques sur les végétaux renfermés dans les diverses couches du globe.* 2 vols. Paris (Dufour and d'Ocagne).
Buch, Leopold von, 1820. "Ueber die Zusammensetzung der basaltischen Inseln und über Erhebungs-Cratere" (Vorgelesen 28 Mai 1818), *Abh. Phys. Klasse, Königl.-Preuss. Akad. Wissensch.*, Berlin, *1818–19*, 51–68.
———, 1825. *Physicalische Beschreibung der Canarischen Inseln.* Berlin (K. Akad. d. Wiss.).
Buckland, William, 1823. *Reliquiae Diluvianae; Or, Observations on the Organic Remains Contained in Caves, Fissures, and Diluvial Gravel, and on Other Geological Phenomena, Attesting the Action of an Universal Deluge.* London (John Murray).
Cordier, [P.] L. [A.], 1827. "Essai sur la température de l'intérieur de la terre" [lu à l'Acad. Sci., 4 juin, 9, 23 juillet 1827], *Mém. Mus. d'Hist. Nat.*, *15*, 161–244; *Mém. Acad. Roy. Sci., Inst. France*, 7 (1827), 473–555.
Cuvier, G., 1812. *Recherches sur les ossemens fossiles de quadrupèdes, ou l'on rétablit les caractères de plusieurs espèces d'animaux que les révolutions du globe paroissent avoir détruites.* 4 vols. Paris (Deterville).
Daubeny, Charles, 1826. *A Description of Active and Extinct Volcanos; With Remarks on their Origin, their Chemical Phaenomena, and the Character of their Products, As Determined by the Condition of the Earth during the Period of their Formation.* London (W. Phillips).
Élie de Beaumont, L., 1829–30. "Recherches sur quelques-unes des révolutions de la surface du globe, présentant différens exemples de coïncidence entre le redressement des couches de certains systèmes de montagnes, et les changemens soudains qui ont produit les lignes de démarcation qu'on observe entre certains étages consécutifs des terrains de sédiment," *Ann. sci. nat.*, *18* (1829), 5–25, 284–416; *19* (1830), 5–99, 177–240.
Fourier, [J. B. J.], 1820. "Extrait d'un mémoire sur le refroidissement séculaire du globe terrestre," *Ann. chim. phys.*, *13*, 418–38.
———, 1827. "Mémoire sur les températures du globe terrestre et des espaces planétaires," *Mém. Acad. Roy. Sci., Inst. France*, 7, 569–604.
Geoffroy St. Hilaire, É., 1825. "Recherches sur l'organisation des gavials; Sur leurs affinités naturelles, desquelles résulte la nécessité d'une autre distribution générique, *Gavialis, Teleosaurus* et *Steneosaurus*; et sur cette question,

si les Gavials (*Gavialis*) aujourd'hui répandus dans les parties orientales de l'Asie, descendent, par voie non interrompue de génération, des Gavials antédiluviens, soit des Gavials fossiles, dits Crocodiles de Caen (*Teleosaurus*), soit des Gavials fossiles du Havre et de Honfleur (*Steneosaurus*)," *Mém. Mus. d'Hist. Nat.*, *12*, 97–155, pls. 5, 6.

————, 1828. "Mémoire où l'on propose de rechercher dans quels rapports de structure organique et de parenté sont entre eux les animaux des âges historiques, et vivant actuellement, et les espèces antédiluviennes et perdues," *Mém. Mus. d'Hist. Nat.*, *17*, 209–29.

Hoff, Karl Ernst Adolf von, 1822–34. *Geschichte der durch Überlieferung nachgewiesenen natürlichen Veränderungen der Erdoberfläche.* 3 vols. Gotha (Justus Perthes).

Hooykaas, R., 1957. "The Parallel between the History of the Earth and the History of the Animal World," *Arch. intern. d'hist. sci.*, no. 38, 3–18.

————, 1959. *Natural Law and Divine Miracle. A Historical-Critical Study of the Principle of Uniformity in Geology, Biology and Theology.* Leiden (E. J. Brill).

Kuhn, Thomas S., 1962. *The Structure of Scientific Revolutions.* Intern. Encycl. Unif. Sci., *2* (2).

Lyell, Charles, 1830–33. *Principles of Geology, Being an Attempt to Explain the Former Changes of the Earth's Surface, by Reference to Causes Now in Operation.* 3 vols. London (John Murray).

Lyell, (Mrs.) [K.], 1881. *Life, Letters and Journals of Sir Charles Lyell, Bart.* 2 vols. London (John Murray).

Scrope, G. Poulett, 1825. *Considerations on Volcanos, the Probable Causes of their Phenomena, the Laws Which Determine their March, the Disposition of their Products, and their Connexion with the Present State and Past History of the Globe; Leading to the Establishment of a New Theory of the Earth.* London (W. Phillips).

————, 1827. *Memoir on the Geology of Central France; Including the Volcanic Formations of Auvergne, the Velay and the Vivarais.* 2 vols. (1 of plates). London (Longman, Rees, Orme, Brown, and Green).

Sedgwick, (Rev.) Adam, Feb. 1831. "Address to the Geological Society, delivered on the Evening of the 18th of February 1831," *Proc. Geol. Soc. Lond.*, *1*, No. 20, 281–316.

Whewell, William, Mar. 1832. [Review of *Principles of Geology*], *Quart. Rev.*, *47*, No. 93, 103–32.

————, 1837. *History of the Inductive Sciences from the Earliest to the Present Time.* 3 vols. London (John W. Parker).

Wilson, Leonard G., 1967. "The Origins of Charles Lyell's Uniformitarianism," *Geol. Soc. Amer. Spec. Pap. 89*, 35–62, 3 pls.

Commentary on the Paper of M. J. S. Rudwick

By Rhoda Rappaport, Vassar College

I AM SO completely in agreement with Dr. Rudwick's general aims, that it is with reluctance that I must disagree with some of his conclusions.

Specifically, to search for the assumptions, the conceptions and preconceptions, of an age not characterized by a single great name or idea or book is an enterprise which historians of geology have too long neglected. To examine searchingly the "isms" which plague and often obscure the history of geology—not only Uniformitarianism and Catastrophism, but also Neptunism, Vulcanism, and Plutonism —is an equally difficult and essential undertaking. We are badly in need of accurate definitions of these terms so that they can be properly used or even, if necessary, discarded. To question Lyell's judgment of his contemporaries and predecessors is especially important since historians have been all too ready to accept those judgments. And finally, to assert that geology was a well developed science by 1830, with established methods and with theories neither fanciful nor speculative, is a point which cannot be made too often or too vigorously.

Less convincing, however, is Dr. Rudwick's thesis that geologists in the 1820's possessed a coherent synthesis to which new evidence and ideas were being accommodated. Directionalism, as he defines it, seems to me to be a cluster of ideas, some of them long familiar and tacitly accepted, others gaining in strength from evidence provided during the 1820's. But the coherence of directionalism as a *synthesis* which was recognized as such during that decade has not, I think, been demonstrated.

Instead, Dr. Rudwick's evidence suggests to me that geologists were less concerned with a general theory than they were with the process of explaining particular phenomena in a fashion which would take into account other, closely related, phenomena. It might thus be more revealing to examine the literature of the *1830's* to see the extent to which the ingredients of directionalism were by then coming to form a coherent whole.

Furthermore, it seems doubtful on chronological grounds that Lyell's *Principles* could have been attacking a directionalist synthesis, although certainly Lyell did attack particular ideas enumerated by Dr. Rudwick; Lyell was, however, developing his own theories during the very years in which the work of Fourier and Cordier was being absorbed by other geologists, and it is thus unlikely that Lyell was responding to a fully formed doctrine.

All this is not to say that geology lacked coherence before 1830, but I think its coherence was of a different order from that suggested by Dr. Rudwick. Geology had long possessed two persistent and unifying traditions which might be summarized as applications of Newton's second Rule of Reasoning in the *Principia*: "Therefore to the same natural effects we must, as far as possible, assign the same causes." Starting from natural effects was a Baconian imperative, while arriving at the same causes clearly implied that nature operates according to uniform, discoverable laws. These two emphases, upon empiricism and the rule of law, characterized the various sciences—and some social philosophies as well—and it is odd that historians have been so reluctant to acknowledge their existence in geology before Lyell's day.

Empiricism in geology sometimes meant the collection of random observations, but it also meant an insistence upon abiding by the evidence in any attempt to formulate theories. Here I think the several eighteenth-century "theories of the earth" have been both over-emphasized and misunderstood. Not only were they far from "typical" of geological writing—they were greatly outnumbered by more specialized treatises and articles—but they were also, in fact, based to a large extent upon the available geological data.

There were still, to be sure, speculations and assumptions, but geologists were self-consciously and piously trying to avoid them, as is shown by their reactions to Buffon's two great works of synthesis; geologists tended to select from his *Théorie de la terre* (1749) and *Les Époques de la nature* (1778) specific information and ideas which might be tested, but they did not fail to criticize the more speculative arguments of Buffon. Empiricism probably gained in strength in the early nineteenth century. As L. Pearce Williams has remarked, referring to the physical scientists of the period, a revival of Baconianism and certain other factors resulted in genuine fear among scientists of appearing to speculate or even of deriving *verifiable* ideas from "impure," speculative sources.

In geology, Baconian orthodoxy was somewhat tempered by a simultaneous religious revival in both England and France; nonetheless, empiricism remained the only proper foundation for geology, and the nineteenth century possessed what Dr. Rudwick has called "a coherent research tradition."

The second Newtonian legacy, an emphasis upon the laws of nature, is too well known for the eighteenth century to require discussion. Scientists of the nineteenth century were perhaps less prone than their predecessors to indulge in professions of faith about Nature, but I find in Dr. Rudwick's casual reference to the recurrence of the phrase, "the ordinary laws of nature," confirmation of the fact that the tradition was alive. And it is this tradition which expresses itself in geological actualism—always with the Newtonian qualification, "as far as possible"—for a period of some fifty years before Lyell.

Concern with empiricism and the rule of law gave geology its peculiar characteristics and problems, and these I can only summarize briefly. First, geologists felt obliged to abide by the evidence, and this meant both geological data and the relevant discoveries and laws of physics and chemistry. Furthermore, explanatory causes were supposed to be both uniform and natural. And the general result was not a coherent synthesis, but confusion, disagreement, and partial syntheses.

The evidence suggested, for example, not that basalt is volcanic, but only that *some* basalts are volcanic in origin. Observations and chemical experiments led to radically different conclusions, depending upon which evidence seemed the more reliable, about the nature of the earth's core. There was increasing evidence for a long geological time scale, but, in Dr. Rudwick's phrase, having "nothing but their geological intuition to go on" led geologists not to speculate but instead to be silent or to offer only the most tentative suggestions. In short, geology possessed a number of partial syntheses, all of them subject to continual debate.

It thus seems to me that empiricism and the rule of law—and the accumulation of new evidence, especially from stratigraphy—continued to be the most salient features of geology through the early decades of the nineteenth century.

These currents provide a more formless picture than that suggested by Dr. Rudwick, but I should like to point out the possibility that it was just this relative formlessness that Lyell attacked, carrying his

attack to dramatic extremes. Geologists had formulated methodological principles, and yet they had not succeeded in arriving at any
one acceptable synthesis or theory; Lyell therefore insisted that
geology lacked a proper method, and his argument was strategically
and logically sound, although it was historically incorrect.

Commentary on the Paper of M. J. S. Rudwick

By Leroy E. Page, Kansas State University

LYELL WROTE in his *Principles of Geology* that the favorite maxim of the Geological Society of London in the early years was that the time was not yet come for a general system of geology, but that all must be content for many years to be exclusively engaged in furnishing materials for future generalizations. I would like to ask whether the time has yet come for a history of geology: a history that will get behind the labels, such as "uniformitarian" and "catastrophist," "Huttonian" and "Wernerian," "Vulcanist" and "Neptunist," to a deeper level of analysis that will recognize the actual diversity and complexity of individuals and beliefs in geology.

I believe that the time has come and that Professor Rudwick's paper is a welcome contribution to such a history. I found it both agreeable and stimulating, although I felt sometimes that he had anticipated all possible objections to his assertions and had chosen his words very carefully with just the right amount of vagueness. He has written an eloquent account that, like Lyell's, all too readily sways its readers.

Let me say that I am not opposed to the use of a label to refer to individuals who have certain beliefs in common. What I object to is thinking a person has been explained when he has been given the proper label. I think that Whewell's terms, "uniformitarian" and "catastrophist," are useful for the period of the controversy over Lyell's views. He defined the terms on several occasions, and I think that the definitions in his *History of the Inductive Sciences* (1837) rather fairly characterize the opposing groups.

He describes the uniformitarians as those who deny any evidence of a beginning or of any material alteration in the energy of earth forces, whereas the catastrophists are those who reject these arbitrary limitations on geological speculation, those who would not prejudge the situation but would allow the evidence to decide the question.

Catastrophists are thus those who will assume catastrophes in the past when the evidence warrants and who believe that geological history consists of one cycle, yet unfinished, rather than a series of uniform cycles. Lyell's argument is that the evidence never does warrant assuming unknown causes because it is always more probable that we are ignorant of all possible effects of existing causes.

Rudwick prefers to call the catastrophists "directionalists," because he considers their directionalism more characteristic than their catastrophism. That is, he assumes the conflict between Lyell's steady-state system and the directionalism of his opponents the more fundamental struggle. This seems, however, not to have been recognized at the time and therefore deserves careful investigation.

Rudwick states that Lyell's fundamental commitment was to the steady-state system. Certainly it was a prior commitment. As early as 1826, Lyell argued in the *Quarterly Review* that past and present-day processes are analogous, but questioned Scrope's conclusion that the power displayed by nature had continually decreased. The greater derangement of the older strata was rather the result, Lyell thought, of the cumulative effect "of the uniform action of the same cause throughout a long succession of ages." The convulsion that produced the Alps was not inferior in violence, he said, to those of the earlier periods. The earth during the period of deposition of the secondary strata was not in a state of chaotic confusion: "There are proofs of occasional convulsions, but there are also proofs of intervening periods of order and tranquillity." He agreed that there was evidence in the past of the destructive action of water over a great extent of the globe, "unparalleled by existing causes," and suggested that these catastrophes might have been related to the sudden elevation or subsidence of land.[1]

Thus we see that Lyell in 1826 was a naturalistic, actualistic, saltatory, anti-directionalist (to use Rudwick's useful terms). This position surely stems from the Hall brand of Huttonianism.

Rudwick mentions that Lyell initially rejected the alleged rise of the Baltic shoreline because it failed to fit his more saltatory theory of elevation by earthquakes. It should be pointed out that he was quick to accept it once he had seen the evidence, and it became a powerful argument for his uniformitarian views. It is true that Lyell was slow to recognize the full potentialities of the long-continued action of streams in large-scale denudation and preferred to employ quicker-acting agents, particularly ocean waves. He was later to be criticized for this by neo-Huttonians.

I don't understand why Rudwick goes on to say that *only* in the conflict between steady-state and directionalism "do we find the characteristic rival interpretations of the same phenomena, the viewing of the evidence through alternative spectacles." Do we not find the same thing in the argument over catastrophes? Doesn't Lyell attribute to many successive earthquakes what his opponents attribute to one gigantic convulsion?

Rudwick discusses four meanings of geological uniformity and their opposites and states that a coherent geological synthesis could be constructed from any combination of these four pairs of categories. He admits that it is sometimes difficult to decide when a cause is so unlike present causes as to merit being called non-actualistic. It was therefore rather easy for Lyell's opponents to protest that Lyell had them confused with somebody else when he implied that they were supranaturalistic non-actualists; and they stoutly maintained that they were really naturalistic actualists, although saltatory directionalists, rather than gradualist steady-staters.

Thus William Conybeare could write that Lyell was mistaken when he suggested that his opponents "speculated on causes of a different order from any with which we are acquainted, and almost reasoned on the supposition of different laws of nature: Whereas I conceive both parties equally ascribe geological effects to known causes, viz. to the action of water, and of volcanic power."[2]

Whewell wrote that, with the exception of the creation of new species, "all the facts of geological observation are *of the same kind* as those which occur in the common history of the world," but perhaps not "*of the same order*,"[3] a distinction identical with that of Rudwick between kind and degree.

Élie de Beaumont, in his paper of 1829, did not seem to relate his epochs of elevation to a specific cause, although they were clearly naturalistic.

Sedgwick, in his presidential address to the Geological Society in 1831, in which he stated the catastrophist position in opposition to Lyell, referred the epochs to "extraordinary volcanic energy" or to "paroxysms of internal energy."[4]

De Beaumont in 1831, although calling the instantaneous elevation of a whole mountain-chain an "event of a different order from those which we daily witness," went on to say that "the mind would not rest satisfied if it did not perceive among those causes *now in action*, an element, fitted from time to time to produce disturbances different from the ordinary march of the phenomena which we now witness . . . the influence exercised by the interior of a planet on its exterior covering during the different stages of refrigeration."[5]

By 1831 de Beaumont clearly regarded his theory as an actualistic one and foresaw the likelihood of future catastrophies. Notice, however, that the supposed cause operates only intermittently and therefore is not strictly observable at the present time. De Beaumont's cause was open to the objection that Lyell had urged against it in 1826, namely that the effect is not proportional to the cause. The magnitude of the upheavals should have declined with time as the earth cooled, whereas in fact the geologically most recent upheavals seem to have been as great as or greater than any of the preceding. Lyell continued to use this argument against the cooling earth theory.

In order to maintain his steady-state theory, Lyell postulated a relationship between a changing distribution of land and sea and world climate, assumed a chemical source of heat within the earth, and revived the Huttonian metamorphic theory to explain the different character of the older rocks, including their lack of fossils. None of these theories were immediately convincing to his friend, Scrope, who also felt that Lyell was wrong in denying the possibility of finding a beginning and in rejecting all catastrophes.[6]

In a letter in 1832, Scrope, in reference to the dispute between Lyell and his opponents, wrote: "I do not see any but an imaginary line of separation between you. It is only a dispute about degree, a plus and minus affair; a little concession on either side will unite you in perfect cordiality. . . . If your antagonists deny the minor degrees of violence altogether, or as being the most frequent, they are decidedly deserting the analogy—but if you deny on your side the probability of the major and catastrophical events having sometimes taken place, you will equally sin against the same law."[7] The analogy he speaks of is one he has made between the probable range of violence of earthquakes and that of volcanoes. One real bond of agreement between Lyell and Scrope at this time was their common desire to free geology, as Scrope put it, "once and forever from the clutches of Moses,"[8] or, using Rudwick's terms, to make geology entirely naturalistic.

In the years immediately after the publication of Lyell's *Principles*, there were a number of statements of the catastrophist-developmentalist synthesis. A rather full presentation of it is to be found in Henry T. De la Beche's *Researches in Theoretical Geology* (1834). In this work, De la Beche wrote of the "millions of years" necessary for the formation of the fossiliferous rocks.[9] While admitting that "it is anything but desirable to have constant recourse to comparatively great power in explaining geological phaenomena when the exertion of a

small force, or of an accumulation of small forces, will afford sufficient explanation of the facts observed," he denied that "an accumulation of small forces will always produce the same effect as the
sudden exertion of a great power equal in intensity to the sum of the
smaller forces."[10]

De la Beche admitted that in many faults it was "exceedingly
difficult to say whether the displacement was produced by one or a
number of repeated shocks" and relied instead upon the widespread
folding in the Alps and elsewhere as his principal evidence for the
exertion in the past of forces of an intensity much greater than that of
a modern earthquake.[11]

Rudwick more or less admits that the synthesis he speaks of was not
well formulated prior to 1830, and yet he calls it a coherent research
tradition. Certainly elements of the synthesis were gaining ground.
For example, Scrope in December, 1828, wrote to Lyell that "Pyro-
geny" was "spreading far and wide." He credited Cordier's *Memoir*
for this, although he didn't regard it as very original. He referred also
to John Herschel, two years before, "broaching the idea of the contraction of the crust of the globe causing the exudation of volcanic
matter," but thought that much more work needed to be done with
this theory before it could be accepted.[12]

William Fitton, speaking in February, 1828, to the Geological
Society, said that "it is no longer denied, that volcanic power has
been active during all the revolutions which the surface of the globe
has undergone, and has probably been itself the cause of many of
them";[13] and in February, 1829, he predicted that "as the doctrine
of Werner, which ascribed to volcanic power an almost accidental
origin, and an unimportant office, has long since expired; so the more
recent views, which regard a certain class of causes as having ceased
from acting, will probably give place to an opinion, that the forces
from whence the present appearances have resulted, are . . . perma-
nently connected with the constitution, and structure of the Globe."[14]

It seems clear to me that Lyell's book stimulated his opponents to
define their position in naturalistic, actualistic terms and that this
position is essentially the synthesis that Rudwick speaks of. Rudwick
has not shown that this synthesis had widespread acceptance, in the
form in which he describes it, prior to Lyell's book.

It is difficult to show that more than a handful of geologists accepted all of it, or even most of it, before 1830. The evidence suggests
to me rather that there arose *in response* to the challenge of Lyell's
synthesis, "another geological synthesis, of equally great scope,
sophistication and explanatory power"; which shared *much* of Lyell's

method, but "differed significantly in its conclusions." Rudwick has done much to clarify our thinking about this synthesis and to defend it from the charge of being unphilosophical or unscientific.

1. *Quarterly Review*, XXXIV (1826), 518, 520.

2. *Philosophical Magazine*, VIII (1830), 359f.

3. *Quarterly Review*, XLVII (1832), 126.

4. *Philosophical Magazine*, IX (1831), 314–15.

5. *Philosophical Magazine*, X (1831), 243, and in a communication quoted by Henry T. De la Beche in his *Geological Manual* (U.S. ed., 1832), 500.

6. Letters to Lyell, June 11, 1830, and May 18, 1833, Darwin-Lyell Correspondence, American Philosophical Society.

7. Letter, March 20, 1832, Darwin-Lyell Correspondence, American Philosophical Society.

8. Letter, June 11, 1830, *op. cit.*

9. *Researches in Theoretical Geology*, 371.

10. *Ibid.*, 123.

11. *Ibid.*, 126, 131–32.

12. Letter, December 23, 1828, Darwin-Lyell Correspondence, American Philosophical Society.

13. *Philosophical Magazine*, III (1828), 295.

14. *Philosophical Magazine*, V (1829), 463.

Archimedes in the Late Middle Ages

By Marshall Clagett, The Institute for Advanced Study

IN SOME RESPECTS the title of my paper is not correct, for in emphasizing the course of Archimedean studies in the late Middle Ages, I have found it necessary to summarize in the beginning the knowledge of Archimedes among the Arabs and the Latin schoolmen in the high Middle Ages to provide some points of contrast and continuity. Furthermore, I have also made numerous references to the fate of the medieval Archimedes in the Renaissance.

Needless to say, the whole paper reflects both my already published first volume on *Archimedes in the Middle Ages: The Arabo-Latin Tradition* and the two subsequent volumes now in preparation. These later volumes will present the full text of the translations of Archimedes made from the Greek by the thirteenth-century Flemish Dominican William of Moerbeke and the step by step use of these translations and the earlier translations from the Arabic down to 1565. Of course, I shall treat here quite lightly many of the problems that I have undertaken to discuss in detail in the three volumes.

This paper also reflects much of the second half of my article on Archimedes that appears in Volume I of the *Dictionary of Scientific Biography*, although I have here often expanded the compendious treatment presented in that article.

Unlike the *Elements* of Euclid, the works of Archimedes were not widely known in antiquity. Our present knowledge of his works depends largely on the interest taken in them at Constantinople from the sixth through the tenth century. It is true that before that time individual works of Archimedes were obviously studied at Alexandria, since Archimedes was often quoted by three eminent mathematicians of Alexandria: Hero, Pappus and Theon. But it is with the activity of Eutocius of Ascalon, who was born toward the end of the fifth century and studied at Alexandria, that the textual history of a collected edition of Archimedes properly begins.

239

Eutocius composed commentaries on three of Archimedes' works: *On the Sphere and the Cylinder, On the Measurement of the Circle,* and *On the Equilibrium of Planes.* These were no doubt the most popular of Archimedes' works at that time.

The Commentary on the Sphere and the Cylinder is a rich work for historical references to Greek geometry. For example, in an extended comment to Book II, Proposition 1, Eutocius presented manifold solutions by earlier geometers to the problem of finding two mean proportionals between two given lines. This commentary also contains the solution of a subsidiary problem, promised by Archimedes for the end of Proposition 4 of Book II of *On the Sphere and the Cylinder* but which was missing from all copies of Archimedes. Eutocius notes that after "unremitting and extensive research" he found such a solution that still retained vestiges of the Doric dialect and the old terminology for conic sections, and which he therefore thought to be Archimedes' solution.[1] Incidentally, this commentary *On the Sphere and the Cylinder* stimulated the composition of a number of tracts and chapters on the problem of finding two mean proportionals in the Middle Ages and the Renaissance, e.g., in the works of Johannes de Muris, Nicholas of Cusa, Leonardo da Vinci, Johann Werner, Francesco Maurolico, Nicholas Tartaglia, and others.

Eutocius' *Commentary on the Measurement of the Circle* is of interest in its detailed expansion of Archimedes' calculation of π. The works of Archimedes and the commentaries of Eutocius were studied and taught by Isidore of Miletus and Anthemius of Tralles, Justinian's architects of *Sancta Sophia* in Constantinople. It was apparently Isidore who was responsible for the first collected edition of at least the three works commented on by Eutocius and his commentaries. Later Byzantine authors seem gradually to have added other works to this first collected edition until the ninth century, when the educational reformer Leon of Thessalonica produced the compilation represented by Greek manuscript A (adopting the designation used by the modern editor, J. L. Heiberg).[2]

Manuscript A contained all of the Greek works of Archimedes now known excepting *On Floating Bodies, On the Method, Stomachion,* and *The Cattle Problem.* This was one of the two Greek manuscripts available to William of Moerbeke when he made his Latin translations in 1269. It was the source, directly or indirectly, of all of the Renaissance copies of Archimedes.

A second Byzantine manuscript, designated as B, included only the mechanical works: *On the Equilibrium of Planes, On the Quadrature of the Parabola*, and *On Floating Bodies* (and possibly *On Spirals*). It too was available to Moerbeke. But it drops out of history after a reference to it in the early fourteenth century (in a Vatican catalogue of 1311).

Finally, we can mention a third Byzantine manuscript, C, a palimpsest whose Archimedean parts are in a hand of the tenth century. It was not available to the Latin West in the Middle Ages, or indeed in modern times until its identification by Heiberg in 1906 at Constantinople (where it had been brought from Jerusalem). It contains large parts of *On the Sphere and the Cylinder*, almost all of *On Spirals*, and some parts of *On the Measurement of the Circle* and *On the Equilibrium of Planes*, and a part of the *Stomachion*. More important, it contains most of the Greek text of *On Floating Bodies* (a text unavailable in Greek since the disappearance of manuscript B) and a great part of the *On the Method of Mechanical Theorems*, hitherto known only by hearsay (Hero mentions it in his *Metrica* and the Byzantine lexicographer Suidas declares that Theodosius wrote a commentary on it).

At about the same time that Archimedes was being studied in Byzantium, he was also finding a place among the Arabs. The Arabic Archimedes has been studied in only a preliminary fashion, but it seems unlikely that the Arabs possessed any manuscript of his works as complete as manuscript A.[3] Still, they often brilliantly exploited the methods of Archimedes and brought to bear their fine knowledge of conic sections on Archimedean problems. The Arabic Archimedes consisted of the following works:

(1) *On the Sphere and the Cylinder* and at least a part of Eutocius' commentary on it. This work seems to have existed in a poor, early translation, revised in the late ninth century, first by Isḥāq ibn Hunain and then by Thābit ibn Qurra. It was re-edited by Naṣīr al-Dīn al-Tūsī in the thirteenth century and was on occasion paraphrased and commented on by other Arabic authors (see the index of Suter's *"Die Mathematiker und Astronomen"* under *Archimedes*).

(2) *On the Measurement of the Circle*, translated by Thābit ibn Qurra and re-edited by al-Tūsī. Perhaps the commentary on it by Eutocius was also translated, for the extended calculation of π found in the ninth-century geometrical tract of the Banū Mūsā bears some resemblance to that present in the commentary of Eutocius.

(3) A fragment of *On Floating Bodies*, consisting of a definition of specific gravity not present in the Greek text, a better version of its

basic postulate than exists in the Greek text, the enunciations without proofs of seven of the nine propositions of Book I and the first proposition of Book II.

(4) Perhaps *On the Quadrature of the Parabola*—at least this problem received the attention of Thābit ibn Qurra.

(5) Some indirect material from *On the Equilibrium of Planes* found in other mechanical works translated into Arabic (such as Hero's *Mechanics*, the so-called Euclid tract *On the Balance*, the *Liber karastonis*, etc.).

(6) In addition, various other works attributed to Archimedes by the Arabs for which there is no extant Greek text: *The Lemmata* or *Liber assumptorum*, *On Water Clocks*, *On Touching Circles*, *On Parallel Lines*, *On Triangles*, *On the Properties of the Right Triangle*, *On Data*, and *On the Division of the Circle into Seven Equal Parts*. Manuscripts of all but two of these works have been noted (and perhaps manuscripts of these two works will also turn up as we study Arabic mathematics in more detail).[4]

Of those additional works, we can single out the *Lemmata* (*Liber assumptorum*), for, although it can not have come directly from Archimedes in its present form, since the name of Archimedes is cited in the proofs, in the opinion of experts several of its propositions are Archimedean in character. One such proposition was Lemma 8, which reduced the problem of the trisection of an angle to a *neusis* or "verging" construction like those used by Archimedes in *On Spiral Lines*.[5]

Special mention should also be made of the *Book on the Division of the Circle into Seven Equal Parts* for its remarkable construction of a regular heptagon that may be originally from Archimedes (its Propositions 16 and 17 lead to that construction).[6] This work stimulated a whole series of Arabic studies of this problem, including one by the famous Alhazen (Ibn al-Haitham).

The key to the whole procedure is an unusual *neusis* presented in Proposition 16 that would allow us to find a straight line divided at two crucial points, that is, a straight line whose divisions allow in Proposition 17 the construction of a circle about the line and the division of its circumference into seven equal arcs. The way in which the *neusis* was solved by Archimedes (or whoever was the author of this tract) is not known. Alhazen, in his later treatment of the heptagon, mentions the Archimedean *neusis* but then goes on to show that one does not need it. Rather he shows that the two crucial points in Proposition 17 can be found by the intersection of a parabola and a hyperbola.[7]

It should be observed that all but two of Propositions 1–13 in this tract concern right triangles, and those two are ones necessary for propositions concerning right triangles. It seems probable, therefore, that Propositions 1–13 comprise the so-called *On the Properties of the Right Triangle* attributed in the *Fihrist* to Archimedes (although at least some of these propositions are Arabic interpolations). Incidentally, Propositions 7–10 have as their objective the formulation $A=(s-a)\cdot(s-c)$, where A is the area and a and c are the sides including the right angle and s is the semiperimeter, and Proposition 13 has as its objective $A=s(s-b)$, where b is the hypotenuse. Hence, if we multiply the two formulations, we have

$$A^2 = s(s-a)\cdot(s-b)\cdot(s-c)$$

or

$$A=\sqrt{s(s-a)\cdot(s-b)\cdot(s-c)},$$

Hero's formula for the area of a triangle in terms of its sides—at least in the case of a right triangle. Interestingly, the Arab scholar al-Bīrūnī attributed the general Heronian formula to Archimedes.

Propositions 14 and 15 of the tract make no reference to Propositions 1–13 and concern chords. Each leads to a formulation in terms of chords equivalent to

$$\sin a/2=\sqrt{(1-\cos a)/2}.$$

Thus Propositions 14–15 seem to be from some other work (and at least Proposition 15 is an Arabic interpolation). If Proposition 14 was in the Greek text translated by Thābit ibn Qurra and does go back to Archimedes, then we would have to conclude that this formula was his discovery rather than Ptolemy's, as it is usually assumed to be.

The Latin West received its knowledge of Archimedes from both the sources just described: Byzantium and Islam. There is no trace of the earlier translations imputed by Cassiodorus to Boethius. Such knowledge that was had in the West before the twelfth century consisted of some rather general hydrostatic information that may have indirectly had its source in Archimedes.

It was in the twelfth century that the translation of Archimedean texts from the Arabic first began.[8] The small tract *On the Measurement of the Circle* was twice translated from the Arabic. The first translation was a rather defective one and was possibly executed by Plato of Tivoli. There are many numerical errors in the extant copies of it and the second half of Proposition 3 on the calculation of π is missing.

The second translation was almost certainly done by the twelfth century's foremost translator, Gerard of Cremona. The Arabic text from which he translated (without doubt the text of Thābit ibn Qurra) included a corollary on the area of a sector of a circle attributed by Hero to Archimedes but missing from our extant Greek text. Not only was Gerard's translation widely quoted by medieval geometers such as Gerard of Brussels, Roger Bacon, Thomas Bradwardine, and others, but it served as the point of departure for a whole series of emended versions and paraphrases of the tract in the course of the thirteenth and fourteenth centuries.

Among these are the so-called Naples, Cambridge, Florence and Gordanus versions of the thirteenth century; and the Corpus Christi, Munich and Albert of Saxony versions of the fourteenth. These versions were expanded by including pertinent references to Euclid and the spelling-out of the geometrical steps only implied in the Archimedean text. In addition, we see attempts to specify the postulates which underlie the proof of Proposition I. For example, in the Cambridge version three postulates (*petitiones*) introduce the text:[9] "[1] There is some curved line equal to any straight line and some straight line to any curved line. [2] Any chord is less than [its] arc. [3] The perimeter of any including figure is greater than the perimeter of the included figure." Furthermore, self-conscious attention was given in some versions to the logical nature of the proof of Proposition I. Thus, the Naples version immediately announced that the proof was to be *per impossibile*, i.e., by reduction to absurdity.

In the Gordanus, Corpus Christi and Munich versions we see a tendency to elaborate the proofs in the manner of scholastic tracts. The culmination of this kind of elaboration appeared in the *Questio de quadratura circuli* of Albert of Saxony, composed some time in the third quarter of the fourteenth century. The Hellenistic mathematical form of the original text was submerged in an intricate scholastic structure which included multiple terminological distinctions and the argument and counter-argument technique represented by initial arguments ("principal reasons") and their final refutations.

Another trend in the later versions was the introduction of rather foolish physical justifications for postulates. In the Corpus Christi version, the second postulate to the effect that a straight line may be equal to a curved line is supported by the statement that "if a hair or silk thread is bent around circumference-wise in a plane surface and then afterwards is extended in a straight line in the same plane, who will doubt—unless he is hare-brained—that the hair or thread is the same whether it is bent circumference-wise or extended in a straight line and is just as long the one time as the other."[10]

Similarly, Albert of Saxony in his *Questio* declared that a sphere can be "cubed" since the contents of a spherical vase can be poured into a cubical vase. A somewhat similar "pouring technique" involving a cylinder and a cube was employed later by Nicholas of Cusa and Francesco Maurolico for the problem of the quadrature of a circle.[11] Incidentally, Albert based his proof of the quadrature of the circle not directly on Proposition X.1 of the *Elements,* as was the case in the other medieval versions of *On the Measurement of the Circle,* but rather on a "betweenness" postulate: "I suppose that with two continuous [and comparable] quantities proposed, a magnitude greater than the 'lesser' can be cut from the 'greater.' "[12] A similar postulate was employed in still another fourteenth-century version of the *De mensura circuli* called the Pseudo-Bradwardine version.

Finally, in regard to the manifold medieval versions of *On the Measurement of the Circle,* it can be noted that the Florence version of Proposition 3 (dateable close to 1400) contained a detailed elaboration of the calculation of π. One might have supposed that the author had consulted Eutocius' commentary, except that his arithmetical procedures differed widely from those used by Eutocius. Furthermore, no translation of Eutocius' commentary appears to have been made before 1450, and the Florence version certainly must be dated before that time. The influence of Gerard's translation was still vividly apparent in a version copied by Regiomontanus in the fifteenth century and in a special work written by Francesco Maurolico in the sixteenth.[13]

In addition to his translation of *On the Measurement of the Circle,* Gerard of Cremona also translated the geometrical *Discourse of the Sons of Moses* (*Verba filiorum*) composed by the ninth-century Arabic mathematicians, the Banū Mūsā. This Latin translation was of particular importance for the introduction of Archimedes into the West. We can single out these contributions of the treatise:

(1) A proof of Proposition I of *On the Measurement of the Circle* somewhat different from that of Archimedes but still fundamentally based on the exhaustion method.

(2) A determination of the value of π drawn from Proposition 3 of the same treatise but with further calculations similar to those found in the commentary of Eutocius.

(3) Hero's theorem for the area of a triangle in terms of its sides (noted above), with the first demonstration of that theorem in Latin (the enunciation of this theorem had already appeared in the writings of the *agrimensores* and in Plato of Tivoli's translation of the *Liber embadorum* of Savasorda).

(4) Theorems for the area and volume of a cone, again with demonstrations.

(5) Theorems for the area and volume of a sphere with demonstrations of an Archimedean character.

(6) A use of the formula for the area of a circle equivalent to $A = \pi r^2$ in addition to the more common Archimedean form, $A = \frac{1}{2}(cr)$. Instead of the modern symbol π the authors used the expression "the quantity which when multiplied by the diameter produces the circumference."

(7) The introduction into the West of the problem of finding two mean proportionals between two given lines. In this treatise we find two solutions: (a) one attributed by the Banū Mūsā to Menelaus and by Eutocius to Archytas, (b) the other presented by the Banū Mūsā as their own but similar to the solution attributed by Eutocius to Plato.

(8) The first solution in Latin of the problem of the trisection of an angle.

(9) A method of approximating cube roots to any desired limit.

The *Verba filiorum* was, then, rich fare for the geometers of the twelfth century when compared with the simplistic geometry of the Roman *agrimensores* or the geometry of Gerbert at the end of the tenth century.

The *Verba filiorum* was quite widely cited in the thirteenth and fourteenth centuries (and indeed it was known to Regiomontanus in the fifteenth century). In the thirteenth century the eminent mathematicians Jordanus de Nemore and Leonardo Fibonacci made use of it. For example, the latter in his *Practica geometrie* excerpted both of the solutions of the mean proportionals problem given by the Banū Mūsā, while the former in his *De triangulis* (if indeed he is the author of this part of the tract) presented one of them together with an entirely different solution, namely that one assigned by Eutocius to Philo of Byzantium. Similarly, Jordanus (or possibly a somewhat later continuator) extracted the solution of the trisection of an angle from the *Verba filiorum* but in addition made the remarkably perspicacious suggestion that the *neusis* can be solved by the use of a proposition from Alhazen's *Optics* which solves a similar *neusis* by conic sections.

The expanded *De triangulis* of Jordanus also contains a solution of the problem of the construction of a regular heptagon which appears to be based on Arabic sources.[14] And incidentally, in view of our interest in the use of these tracts later, we can note that Regiomontanus read the *De triangulis* of Jordanus[15] and that much of

Leonardo Fibonacci's *Practica* reappears in the *Summa de arithmetica etc.* of Luca Pacioli (Venice, 1494), a work of some influence in the Renaissance.[16]

Some of the results and techniques of *On the Sphere and the Cylinder* also became known through a treatise entitled *De curvis superficiebus Archimenidis* and said to be by Johannes de Tinemue. This seems to have been translated from the Greek in the early thirteenth century or at least composed on the basis of a Greek tract. The *De curvis superficiebus* contained ten propositions with several corollaries and was concerned for the most part with the surfaces and volumes of cones, cylinders and spheres. This work is essentially Archimedean, employing a version of the method of exhaustion. It simplifies the proof of Archimedes' main conclusions by assuming that to any plane surface there exists an equal conical, cylindrical or spherical surface; and that with two surfaces given there is a surface of the same kind as and symmetrically akin to one of the given surfaces and equal to the other.

These assumptions are coupled with the principle that an "included figure" cannot be greater than an "including figure." The including figure for this author always surrounds the included and in no way touches it (cf. Euclid, *Elements*, Proposition XII.16). The method that the author used was probably suggested to him by the proof of Proposition XII.18 of the *Elements* of Euclid.[17]

The *De curvis superficiebus* was a very popular work and was often cited by later authors. Like Gerard of Cremona's translation of *On the Measurement of the Circle*, the *De curvis superficiebus* was emended by Latin authors, two original propositions being added to one version (represented by manuscript *D* of the *De curvis superficiebus*) and three quite different propositions being added to another (represented by manuscript *M* of the *De curvis*). In the first of the additions to the latter version, the Latin author applied the exhaustion method to a problem involving the surface of a segment of a sphere, showing that at least this author had made the method his own. The techniques and propositions of the *De curvis superficiebus* were taken over completely by Francesco Maurolico and integrated beautifully with Archimedes' *On the Sphere and the Cylinder* in 1534.[18]

About the same time as the appearance of the *De curvis superficiebus* in the early thirteenth century, the geometer, Gerard of Brussels, in his *De motu* used Archimedean *reductio* techniques in a highly original manner. He compared two rotating figures by comparing the motions of two corresponding line elements of the figures. This has, as I have pointed out, some similarity with the method used by Archimedes of balancing corresponding line or surface elements of two figures.[19] But Gerard's methods, while owing something to Archimedean geometry, appear to be essentially his own.

In 1269, some decades after the appearance of the *De curvis superficiebus*, the next important step was taken in the passage of Archimedes to the West when much of the Byzantine corpus was translated from the Greek by the Flemish Dominican, William of Moerbeke. In this translation Moerbeke employed Greek manuscripts A and B which had passed to the Pope's library in 1266 from the collection of the Norman Kings of the Two Sicilies that Charles of Anjou gave to the Pope after Manfred's defeat at Benevento. All the works included in manuscripts A and B except for *The Sandreckoner* and Eutocius' *Commentary on the Measurement of the Circle* were rendered into Latin by William. Needless to say, *On the Method*, *The Cattle-Problem*, and the *Stomachion*, all absent from manuscripts A and B, were not among William's translations. Although William's translations are not without error (and indeed some of the errors are serious),[20] the translations, on the whole, present the Archimedean works in an understandable if literal way.

We possess the original holograph of Moerbeke's translations in manuscript Vatican, Ottobonianus latinus 1850. This manuscript was not widely copied. The translation of *On Spirals* was copied from it in the fourteenth century (MS Vat. Reg. lat 1253, 14r–33r); several works were copied from it in the fifteenth century in an Italian manuscript now at Madrid (Bibl. Nac. 9119), and one work (*On Floating Bodies*) was copied from it in the sixteenth century (Vat. Barb. lat. 304, 124r–41v, 160v–61v).[21]

But, in fact, the Moerbeke translations were utilized more than one would expect from the paucity of manuscripts. They were used by several schoolmen at the University of Paris toward the middle of the fourteenth century. Chief among them was the astronomer and mathematician Johannes de Muris, who appears to have been the compositor of a hybrid tract in 1340 entitled *Circuli quadratura*.[22]

This tract consisted of fourteen propositions. The first thirteen were drawn from Moerbeke's translation of *On Spirals* and were just those propositions necessary for the proof of Proposition 18 of *On Spirals*: "If a straight line is tangent to the extremity of a spiral described in the first revolution, and if from the point of origin of the spiral one erects a perpendicular on the initial line of revolution, the perpendicular will meet the tangent so that the line intercepted between the tangent and the origin of the spiral will be equal to the circumference of the first circle." The fourteenth proposition of the hybrid tract was Proposition 1 from Moerbeke's translation of *On the Measurement of the Circle*. Thus this author realized that by the use of Proposition 18 from *On Spirals*, he had achieved the necessary rectification of the circumference of a circle preparatory to the final quadrature of the circle accomplished in *On the Measurement of the Circle*, Proposition 1.

Incidentally, the hybrid tract did not merely use the Moerbeke translations verbatim but also included considerable commentary. In fact, this medieval Latin tract was the first known commentary on Archimedes' *On Spirals*. That the commentary was at times quite perceptive is indicated by the fact that the author suggested that the *neusis* introduced by Archimedes in Proposition 7 of *On Spirals* could be solved by means of an *instrumentum conchoydeale*.[23] The only place in which a medieval Latin commentator could have learned of such an instrument would have been in that section of Eutocius' *Commentary on the Sphere and the Cylinder*, Book II, Proposition 1, where Eutocius describes Nicomedes' solution of the problem of finding two mean proportionals.

We have further evidence that Johannes de Muris knew of Eutocius' *Commentary* in the Moerbeke translation when he used sections from this commentary in his *De arte mensurandi* (Chapter VIII Proposition 16) where three of the solutions of the mean proportionals problem given by Eutocius are presented, namely the solutions attributed to Plato, Hero, and Philo.[24] Not only did Johannes incorporate the whole hybrid tract *Circuli quadratura* into Chapter VIII of his *De arte mensurandi* (composed, it seems, shortly after 1343) but in Chapter X of the *De arte* he quoted verbatim many propositions from Moerbeke's translations of *On the Sphere and the Cylinder* and *On Conoids and Spheroids* (which latter he misapplied to problems concerning solids generated by the rotation of circular segments).

Incidentally, the treatment of spiral lines found in the *Circuli quadratura* seems to have influenced the accounts of spirals found in the *De trigono balistario* completed by Giovanni da Fontana in 1440 (MS Bodleian, Canon. Misc. 47, 216v–19v) and in the *Quadratura circuli* of Nicholas of Cusa, dated December, 1450.[25]

Within the next decade or so after Johannes de Muris, his colleague at the University of Paris, Nicole Oresme, in his *De configurationibus qualitatum et motuum* (Part I, Chapter 21) revealed knowledge of *On Spirals*, at least in the form of the hybrid *Circuli quadratura*.[26] His description of the spiral as an example of uniformly difform curvature is reflected in a similar description given by Leonardo da Vinci (Institut MS *E*, 34v). Further, Oresme in his *Questiones super de celo et mundo* quoted at length from Moerbeke's translation of *On Floating Bodies*, while Henry of Hesse, Oresme's junior contemporary at Paris, quoted briefly therefrom.[27]

Before this time, the only knowledge of *On Floating Bodies* had come in a thirteenth-century treatise entitled *De ponderibus Archimenidis sive de incidentibus in humidum*, a pseudo-Archimedean treatise prepared on the basis of Arabic sources. Its first proposition expressed the basic conclusion of the "principle of Archimedes": "The weight of any body in air exceeds its weight in water by the weight of a volume of water equal to its own volume."[28]

The references by Oresme to the genuine *On Floating Bodies* comprise citations to Propositions 3–7 of Book I and Proposition 1 of Book II. The whole section is particularly interesting because Oresme joins the dynamic definition of specific weight (perhaps derived from the *Liber de ponderoso et levi*, a text that appeared in the thirteenth century in Latin)[29] with the Archimedean considerations of *On Floating Bodies*. It is just such a juxtaposition that appears in Tartaglia's Italian translation and commentary on Book I of *On Floating Bodies*,[30] which may have led Benedetti to his modified form of the Peripatetic law, namely that bodies fall with a speed proportional to the excess in specific weight of the falling body over the medium.[31]

Incidentally, most scholars do not realize that Leon Battista Alberti made observations much like Tartaglia a century earlier. Such observations were not tied directly to the text of *On Floating Bodies*,[32] although the substance of some of the Archimedean propositions is given.

Returning to the fourteenth century, we can conclude from our previous discussion that incontrovertible evidence shows that at the University of Paris in the mid-fourteenth century six of the nine Archimedean translations of William of Moerbeke were known and used: *On Spirals*, *On the Measurement of the Circle*, *On the Sphere and the Cylinder*, *On Conoids and Spheroids*, *On Floating Bodies*, and Eutocius' *Commentary on the Sphere and the Cylinder*. While no direct evidence exists of the use of the remaining three translations, there has been recently discovered in a manuscript written at Paris in the fourteenth century (BN lat. 7377B, 93v–94r) an Archimedean-type proof of the law of the lever that might have been inspired by Archimedes' *On the Equilibrium of Planes*.[33] But other than this, the influence of Archimedes on medieval statics was entirely indirect.

The anonymous *De canonio*, translated from the Greek in the early thirteenth century, and Thābit ibn Qurra's *Liber karastonis*, translated from the Arabic by Gerard of Cremona, passed on this indirect influence of Archimedes in three respects.[34]

(1) Both tracts illustrated the Archimedean type of geometrical demonstrations of statical theorems and the geometrical form implied in weightless beams and weights that were really only geometrical magnitudes.

(2) They gave specific reference in geometrical language to the law of the lever (and in the *De canonio* the law of the lever is connected directly to Archimedes).

(3) They indirectly reflected the centers-of-gravity doctrine so important to Archimedes, in that both treatises employed the practice of substituting for a material beam segment a weight equal in weight to the material segment but hung from the middle point of the weightless segment used to replace the material segment.

Needless to say, these two tracts played an important role in stimulating the rather impressive statics associated with the name of Jordanus de Nemore.

In the fifteenth century, knowledge of Archimedes in Europe began to expand. While the medieval texts continued to influence various authors of the fifteenth century, as I have indicated above,[35] a new source for Archimedes appeared, namely the Latin translation made by Jacobus Cremonensis in about 1450 by order of Pope Nicholas V. Since this translation was made exclusively from Greek manuscript A, the translation failed to include *On Floating Bodies*, but it did include the two treatises in A omitted by Moerbeke, namely *The Sandreckoner* and Eutocius' *Commentary on the Measurement of the Circle*.

It appears that this new translation was made with an eye on Moerbeke's translations.[36] Not long after its completion, a copy of the new translation was sent by the Pope to Nicholas of Cusa, who made some use of it in his *De mathematicis complementis*, composed in 1453–54; although to be sure his earlier works show some smattering of knowledge from medieval sources of *On Spiral Lines, On the Measurement of the Circle*, and *On the Sphere and the Cylinder*, as well as of Eutocius' commentary on the last of these works. There are at least nine extant manuscripts of this new translation, one of which was corrected by Regiomontanus and brought to Germany about 1468 (the Latin translation published with the *editio princeps* of the Greek text in 1544 was taken from this copy). Leonardo da Vinci appears to have seen copies of both the Moerbeke and the Cremonensis translations, and both translations can be shown to have exerted some influence on him.[37]

Incidentally, the fate of Moerbeke's holograph copy of his translations in the fourteenth and fifteenth centuries is not known with any exactness, although we can speculate on its possible transfer to France in the fourteenth and its reappearance in Italy, in the fifteenth century, possibly in the hands of Paolo Toscanelli. As I have said, it would seem to have been seen by Jacobus Cremonensis when he made his new translation. At some unspecified time it appears to have passed into the possession of Pietro Barozzi, Bishop of Padua from 1488 to 1507.[38] At any rate, in the latter date it was acquired by Andreas Coner in Padua and he emended it thoroughly both by comparing it to a Greek manuscript and paying close attention to mathematical sense. From the time of Coner we can easily trace its history until its acquisition by the Vatican Library as part of the Ottobonian collection.

Greek manuscript A itself was copied a number of times. Cardinal Bessarion had one copy prepared between 1449 and 1468 (MS E). Another (MS D) was made from A when it was in the possession of the well-known humanist George Valla. The fate of A and its various copies has been traced skillfully by J. L. Heiberg in his edition of Archimedes' *Opera*. The last known use of manuscript A occurred in 1544, after which time it seems to have disappeared.

The first printed Archimedean materials were in fact merely Latin excerpts of Book II, Propositions 1 and 4, of *On the Sphere and the Cylinder* and Eutocius' lengthy comments on those propositions, that appeared in George Valla's *De expetendis et fugiendis rebus opus* (Venice, 1501) and were based on his reading of manuscript A, which was then in his possession.[39] But the earliest actual printed texts of Archimedes were the Moerbeke translations of *On the Measurement of the Circle* and *On the Quadrature of the Parabola*, published on the basis of the Madrid manuscript in Venice, 1503, by L. Gaurico (*Tetragonismus, id est circuli quadratura etc.*).

In 1543, also at Venice, N. Tartaglia republished the same two translations directly from Gaurico's work, and, in addition, from the same Madrid manuscript, the Moerbeke translations of *On the Equilibrium of Planes* and Book I of *On Floating Bodies* (leaving the erroneous impression that he had made these translations from a Greek manuscript, which he had not since he merely repeated the texts of the Madrid manuscript with virtually all their errors). Incidentally, Curtius Troianus, in Venice, 1565, published from the legacy of Tartaglia both books of *On Floating Bodies* in Moerbeke's translation.

The key event, however, in the further spread of Archimedes was the aforementioned *editio princeps* of the Greek text with the accompanying Latin translation of Jacobus Cremonensis at Basel in 1544. It was printed from a Nürnberg manuscript which had been copied from Greek MS A and contains corrections introduced from the Latin translation of Moerbeke. Since the Greek text rested ultimately on manuscript A, the *On Floating Bodies* was not included.

A further Latin translation of some of the Archimedean texts was published by the perceptive mathematician Federigo Commandino in Bologna in 1558, which the translator supplemented with a skillful mathematical emendation of Moerbeke's translation of *On Floating Bodies* (Bologna, 1565), without any knowledge of the long-lost Greek text. Already in the period 1534–49, a paraphrase of Archimedean texts with some attention to medieval sources had been made by Francesco Maurolico, as I have already noted. This was published in Palermo in 1685 on the basis of an incomplete edition of some years earlier that had been abandoned before general publication.

One other Latin translation by Antonius de Albertis, completed in the early sixteenth century, remains in manuscript only and appears to have exerted no influence on sixteenth-century mathematics and science.

After 1544 the publications on Archimedes and the use of his works began to multiply markedly. His works presented quadrature problems and propositions that mathematicians sought to solve and demonstrate not only with his methods, but also with a developing geometry of infinitesimals that was to anticipate in some respect the infinitesimal calculus of Newton and Leibniz. His hydrostatic conceptions were used to modify Aristotelian mechanics. Archimedes' influence on mechanics and mathematics can be seen in the works of such authors as Commandino, Guido Ubaldi del Monte, Benedetti, Simon Stevin, Luca Valerio, Kepler, Galileo, Cavalieri, Torricelli, and numerous others. For example, Galileo mentions Archimedes more than a hundred times, and the limited inertial doctrine used in his analysis of the parabolic path of a projectile is presented as an Archimedean-type abstraction.

Archimedes began to appear in the vernacular languages. Tartaglia had already rendered into Italian Book I of *On Floating Bodies*, Book I of *On the Sphere and the Cylinder* (apparently on the basis of the translation of Moerbeke) and the section on proportional means from Eutocius' *Commentary on the Sphere and the Cylinder*.

Book I of *On the Equilibrium of Planes* was translated into French in 1565 by Pierre Forcadel. In the same year he published a translation of the medieval *De ponderibus Archimenidis* (or *De incidentibus in humidum*).

It was, however, not until 1670 that a more or less complete translation of Archimedes' works was made into German by J. C. Sturm on the basis of the influential Greek and Latin edition of David Rivault, Paris, 1615. Also notable for its influence was the new Latin edition of Isaac Barrow (London, 1675).

Of the many editions prior to the modern edition of Heiberg, the most important was that of Joseph Torelli, published at Oxford in 1792. By this time, of course, Archimedes' works had been almost completely absorbed into European mathematics and had exerted their not inconsiderable influence on early modern science.

1. Archimedes, *Opera omnia*, ed. J. L. Heiberg, Vol. 3, 130–32.

2. *Ibid.*, xxii. The reference to Leon is given by Heiberg from Greek MS G and he reasons that it was taken from MS A. My account of the Greek manuscripts is based almost entirely on Heiberg's Prolegomena to Vol. 3 of his edition of Archimedes.

3. I have noted the principal literature and known manuscripts of the Arabic Archimedes in the bibliography to my article on Archimedes in the *Dictionary of Scientific Biography*.

4. The only works of which I have found no reference in Arabic manuscripts are those *On Parallel Lines* and *On Data* but I have not made any extensive search of Arabic collections which may well contain these works.

5. For an English translation and discussion of this proposition of the *Lemmata*, see my *Archimedes*, Vol. 1, 667–68.

6. A German translation of these propositions has been given by C. Schoy, *Die trigonometrischen Lehren des persischen Astronomen Abu 'l-Raihân Muh. ibn Ahmad al-Bîrûnî* (Hanover, 1927), 74–84. The whole work has been analyzed in modern fashion by J. Tropfke in *Osiris*, Vol. 1 (1936), 636–51. I have given an English translation of Propositions 16 and 17 in my article on Archimedes in the *DSB*.

7. Schoy, *op. cit.* in note 6, pp. 85–91.

8. Much of the succeeding account of the Arabo-Latin Archimedes is taken from my *Archimedes in the Middle Ages*, Vol. 1: *The Arabo-Latin Tradition* (Madison, 1964).

9. *Ibid.*, 69.

10. *Ibid.*, 170–71.

11. Strictly speaking, the method of Cusa is not a pouring technique but a weighing technique. His method, as described in the *De staticis experimentis* (Part IV of the *Idiota*), which I have read in the edition of L. Baur (Leipzig, 1937), 138, is to take a cylinder (whose base is the circle to be squared) with a given height and then a cube of the same height (with a base equal to the square of the diameter). Each of these vessels is filled with water and weighed. Hence the ratio of the circle to the square is as the ratio of the weights.

Maurolico's procedure does involve pouring. He takes a cylinder with height equal to the diameter of its base and a cube whose edge is equal to same diameter. (*Admirandi Archimedis Syracusani monumenta omnia mathematica, quae extant . . . ex traditione . . . D. Francisci Marolici*, Palermo, 1685, p. 39; and see Paris MS, Bibl. Nat. 7464, f. 25r). Maurolico then suggests filling the cylinder to the top. Afterwards, he pours the contents of the cylinder into the cube and notes the height to which it rises. This permits him to show by geometry that the circle is equal to a rectangle comprised by the height of the liquid in the cube and the diameter of the cylindrical base. He then suggests converting that rectangle into a square by finding the mean proportional between these quantities. The resulting square will be equal to the circle.

Somewhat later in the century John Dee in his preface to H. Billingsley's English translation of Euclid's *Elements* (London, 1570), sig. c i verso, suggests making a sphere and a cube with edge equal to the diameter of the sphere of the same uniform material and then weighing them to show that the ratio of their volumes is 11 to 21. He notes that the ratios of other volumes to each other can be determined in the same way. John Dee was very widely acquainted with medieval mathematical texts.

12. Clagett, *Archimedes*, Vol. 1, 418–19.

13. The version of Regiomontanus occurs in Vienna, Nat.-bibl. 5203, 131v-33r, written by Regiomontanus between 1454 and 1462. It is a paraphrase of the Gordanus Version which I published in *Archimedes*, Vol. 1, 142–65. For Maurolico's version, see his *Archimedis de circuli dimensione libellus*, published in the collected edition given in footnote 11. I have prepared the text and an English translation of this from the holograph of Maurolico (Paris, Bibl. Nat. lat. 7465, 21v–28v).

14. *De triangulis libri IV*, ed. of M. Curtze (Thorn, 1887), 42–44. I have collected films of all of the known manuscripts of the *De triangulis* and there is some evidence that there was an earlier edition without Propositions IV.12–IV.28. These latter propositions, which contain among others the interesting propositions on the finding of two mean proportional, the trisection of an angle, and the construction of a regular heptagon, appear to have circulated separately. I intend to study the text in detail in a later volume.

15. See E. Zinner, *Leben und Wirken des Johannes Müller von Königsberg, genannt Regiomontanus*, 2nd ed. (Osnabrück, 1968), 318.

16. The pertinent parts of Pacioli's *Summa* will be discussed in Vol. 3 of my *Archimedes*.

17. As I have indicated below, the techniques and propositions of the *De curvis superficiebus* were used by Francesco Maurolico in his version of *On the Sphere and the Cylinder* (*ed. cit.* in footnote 11, pp. 40–85). Without naming this treatise, Francesco calls the method "an easier way." (*Ibid.*, 2)

Incidentally, Legendre adapts the method of Proposition XII.18 of the *Elements* to the proof of Proposition XII.2 in the same way as the author of *De curvis superficiebus* does in his somewhat similar Proposition III. For Legendre's proof, see Euclid, *The Elements*, translated with introduction and commentary by Thomas Heath, Vol. 3 (Annapolis, 1947), 377–78, 434–37. For the proof in the *De curvis superficiebus*, see my *Archimedes*, Vol. 1, 462–67. Furthermore, the Italian mathematician, V. Flauti, without realizing that Maurolico or Johannes de Tinemue had employed this method, used it in preparing his *Corso di geometria elementare* in 1808. He followed it for the proof of other propositions of Book XII of the *Elements* as well as for theorems of *On the Sphere and the Cylinder* (V. Flauti, "Sull' Archimede e l'Apollonio di Maurolico," *Memorie della Reale Accademia delle scienze dal 1852 in avanti ripartite nelle tre classi di matematiche, scienze naturali, e scienze morale*, Vol. II (Napoli, 1857), p. XCIII.)

As Flauti notes, he was much later to discover Maurolico's use of this method. The *De curvis superficiebus* was unknown to Flauti and so he did not realize that Maurolico had a predecessor in applying this method to *On the Sphere and the Cylinder*.

18. I shall republish Maurolico's text of *On the Sphere and the Cylinder* with a close analysis of it and a comparison of it with the *De curvis superficiebus* in Vol. 3 of my *Archimedes*. Although Maurolico's Archimedean works were not published until 1685, their composition had begun as early as 1534 and was completed in 1550; the dates of completion of the various works are noted in the 1685 edition.

19. See my *Archimedes*, Vol. 1, 9–10.

20. A partial list of William's errors and misunderstandings of the Greek text was given by Heiberg, *Archimedis opera omnia*, Vol. 3, li–lii. To these I shall add a number of other examples in Vol. 2 of my *Archimedes in the Middle Ages*. It is of interest that in *On the Equilibrium of Planes* Moerbeke often mistranslates τόμοσ by *sector*, when the meaning is rather that of frustum. A possible explanation for this is that several times the Greek text mistakenly has τομεύσ which indeed ought to be rendered by *sector*. Of course the context should have suggested even in these cases that *sector* was not the proper rendering.

Interestingly enough, when he later came to translate *On Conoids and Spheroids*, Moerbeke abandoned the erroneous translation and merely retained the word *tomos* in his Latin text. Moerbeke was also troubled by the Greek word *helix* (i.e., spiral) when he began to translate *On Spirals* and he used words like *volutio* and *revolutio* until settling down with *elix* or the adjectival form *elicus*. The reason for this is not hard to understand since *helix* was commonly used in Latin for ivy or a twisting vine and certainly Moerbeke had no experience with a mathematical *helix* before seeing Archimedes' treatise.

Moerbeke was also greatly puzzled by the signs for fractions and myriads used in Proposition 3 of *On the Measurement of the Circle* and he bungles the large numbers and fractions used in the calculation of π. One would suppose, therefore, that he was unacquainted with the various versions in the Arabo-Latin tradition of *On the Measurement of the Circle*.

21. Heiberg did not know of this last manuscript. Hence he concluded that, since Commandino in making his emended version of *On Floating Bodies* had not used the Madrid manuscript, he must have used Moerbeke's holograph. However, I am reasonably sure that it was Vat. Barb. lat. 304 which Commandino used. This is made virtually certain by the fact that the manuscript includes as the only other item the *De analemmate* of Ptolemy in Moerbeke's translation which Commandino also revised.

22. This treatise will be edited and published for the first time in my *Archimedes*, Vol. 3. I shall also demonstrate the likelihood of Johannes de Muris' authorship.

23. Also perceptive was the realization on the part of the author of this commentary that Archimedes' intention in presenting these *neuseis* in the *On Spirals* was merely to assert the existence of a solution rather than to suggest what the solution was.

24. M. Clagett, "Johannes de Muris and the Problem of Proportional Means," *Medicine, Science and Culture: Historical Essays in Honor of Owsei Temkin*, ed. by L. G. Stevenson and R. Multhauf (Baltimore, 1968), 35–49. In my *Archimedes*, Vol. 3, I shall present the full texts and translations of all of the Archimedean sections of Johannes de Muris' *De arte mensurandi*. I shall also show that Johannes de Muris took an incomplete version of the *De arte mensurandi* (containing only five chapters) and expanded it in an impressive way to include the full range of medieval geometric knowledge.

25. Published with Regiomontanus' *De triangulis* (Norimbergae, 1533), p. 5 separate pagination.

26. M. Clagett, *Nicole Oresme and the Medieval Geometry of Qualities and Motions* (Madison, Wisc., 1968), 220–23. Oresme made use of the spiral in an exceedingly fertile section on the measure of curvature. He represented its curvature by a right triangle after plotting the radius lengths against the values of the angle of rotation (*Ibid.*, 450).

27. The citations to Archimedes' *On Floating Bodies* by Oresme occur in his *Questiones super de celo et mundo*, ed. and transl. by C. Kren (Thesis, Univ. of Wisconsin, 1965), 847–64. The reference by Henry of Hesse is found in his *Questiones super communem perspectivam*, MS Erfurt, Stadtbibl. Amplon. F. 380, 30v, col. 1. The pertinent passages from both of these works will be given in my *Archimedes*, Vol. 3.

28. M. Clagett, *The Science of Mechanics in the Middle Ages* (Madison, Wisc., 1959), 95.

29. *Ibid.*, 435. Albert of Saxony, Oresme's junior contemporary at Paris, gives the dynamic definition of specific weight neatly (*Ibid.*, 137): "With two solid bodies given, it is possible without weighing [them] in a balance to find out [1] whether they are of the same or of different specific weights and [2] which of them is heavier. For let these two bodies be *a* and *b* and let equal volumes of *a* and *b* be taken . . . ; and let these portions be released so that they fall in the same water. Then if they descend equally fast toward the bottom of the water, or if just as large a part of one as of the other [is submerged in the water], say that the said bodies are of equal specific weight. If, however, they descend to the bottom unequally fast, or more of one of these portions is submerged in the water and less of the other, say that the one is heavier according to species whose portion descends more quickly or whose portion is submerged further in the water."

This passage is reflective of the much longer treatment by Oresme referred to in footnote 27, but there is no specific reference here to Archimedes' *On Floating Bodies*.

30. See Tartaglia's *Ragionamenti . . . sopra la sua travagliata inventione* (Venice, 1551) with pagination, but see the second page: "*Et che quelli corpi solidi che sono poi di natura piu gravi di l'acqua posti che siano in acqua, subito se fanno dar loco alla detta acqua, e che non solamente intrano totalmente in quella, ma vanno discendendo continuamente per fin al fondo, e che tanto piu velocemente vanno discendendo quanto che sono piu gravi dell' acqua.*"

31. G. B. Benedetti in his preface to the *Resolutio omnium Euclidis prob-*

lematum etc. (Venice, 1553), sig. ** verso, notes that Tartaglia taught him the first four books of Euclid. He also gives the first exposition of his views of the variation of velocity with specific weight. The main conclusion (sig. *** recto) is: *"Modo dico quod si fuerint duo corpora, eiusdem formae, eiusdemque speciei, aequalia invicem, vel inaequalia, per aequale spacium, in eodem medio, in aequali tempore ferentur."* He had earlier mentioned Archimedes. Cf. my *The Science of Mechanics,* p. 665.

It may be noted that John Dee in his preface to H. Billingsley's translation: *The Elements of Geometrie of . . . Euclide of Megara* (London, 1570), sig. b iiii verso to c i recto, gives a translation of some of the propositions from Archimedes' *On Floating Bodies* and notes that by these propositions "great Errors may be reformed, in Opinion of the Naturall Motion of things, Light and Heavy, Which errors, are in Naturall Philosophie (almost) of all men allowed: too much trusting to Authority: and false Suppositions. As, 'Of any two bodyes, the heavyer, to move downward faster than the lighter.' This error, is not first by me Noted: but by one Iohn Baptist de Benedictis. The chief of his propositions, is this: which seemeth a Paradox. 'If there be two bodyes of one forme, and of one kynde, aequall in quantitie or unaequall, they will move by aequall space, in aequall tyme: So that both theyr movynges be in ayre, or both in water: or in any one Middle [i.e. Medium].' " [*Single quotation marks mine.*]

32. Alberti in his *De' Ludi matematici,* Chap. XX (*Opera volgari,* ed. of A. Bonucci, Vol. 4 [Florence, 1847], 438–39), takes up the crown problem, gives the essential content of Propositions 5–7 of Book I of *On Floating Bodies* and concludes with the dynamic definition: *"E quelli corpi che in sè pesano più che l'acqua staranno sotto; e quanto più peseranno tanto più veloci descenderanno e meno occuperanno dell' acqua sendo tutti d'una figura e forma."* And so once more Tartaglia's ideas seem to be derivative in character.

33. M. Clagett, "A Medieval Archimedean-type Proof of the Law of the Lever," *Miscellanea André Combes,* Vol. II (Rome, 1967), 409–21.

34. See my *Archimedes,* Vol. 1, 9.

35. We have mentioned influences on Giovanni Fontana, Nicholas of Cusa, Leon Battista Alberti, Luca Pacioli and Leonardo da Vinci. We can also mention the influence of the Arabo-Latin tradition of Archimedes on the *Artis metrice practice compilatio* of Leonardo de Antoniis, a Franciscan who composed some geometrical notes on Campanus about 1404/5 in Bologna.

The Italian translation of the *Compilatio* was edited by M. Curtze in the *Abhandlungen zur Gesch. d. Math.,* 13. Heft (1902), 339–434. Curtze misidentified the author with Leonardo Mainardi (fl. 1488). The correct identification was given by A. Favaro in *Bibliotheca Mathematica,* 3. Folge, Vol. 5 (1904), 326–41; cf. his article in *Atti R. Intituto Veneto,* Vol. 63 (1904), 377–95. I am giving an edition of the Archimedean parts of this tract based on several Latin manuscripts in my *Archimedes,* Vol. 3.

36. J. L. Heiberg, "Neue Studien zu Archimedes," *Abhandlungen zur Geschichte der Mathematik 5.* Heft (1890), 83–84, concluded that Jacobus Cremonensis had made use of Moerbeke's translation.

My own study of the Cremonensis translation leads me to conclude that occasionally Cremonensis did indeed consult the Moerbeke translation. While Cremonensis did give better translations than Moerbeke in a number of cases, in one spectacular case Moerbeke's version was better; see S. Heller, "Ein Fehler in einer Archimedes-Ausgabe, seine Entstehung und seine Folgen," *Abhandlungen der Bayerischen Akademie der Wissenschaften. Mathematisch-naturwissenschaftliche Klasse, Neue Folge*, 63. Heft (1954), 21.

Incidentally, Heller believed that Jacobus had not known the work of Moerbeke (*Ibid.*, 19). But all his evidence shows is that Jacobus did not pay much attention to the earlier work in certain given passages, which is certainly true.

37. The first important (but quite incomplete) study of the relation between Leonardo and Archimedes was made by A. Favaro, "Archimede e Leonardo da Vinci," *Atti del Reale Instituto Veneto di Scienze, Lettere ed Arti*, Vol. 71 (1911–12), 953–75. Also incomplete but more interesting are the various sections on Archimedes and Leonardo in R. Marcolongo, *Studi Vinciani: Memorie sulla geometria e la meccanica di Leonardo da Vinci* (Naples, 1937).

I have prepared for Volume 3 of my *Archimedes* a new and detailed study of all the relevant passages in Leonardo's notebooks that may reflect some direct knowledge of Archimedean texts. I have shown some traces in Leonardo's notebooks of the following Archimedean works: (1) *On the Measurement of the Circle*, (2) *On Spirals*, (3) *On the Sphere and the Cylinder*, (4) *On the Equilibrium of Planes*, (5) *On Floating Bodies*, and perhaps (6) Eutocius' *Commentary on the Sphere and the Cylinder*.

Although Leonardo mentions two manuscripts of Archimedes—one apparently of the Moerbeke translation and the other of the Cremonensis translation—some of his knowledge (particularly of works 1, 2 and 3) seems to be second-hand and based on medieval works. Most important was the influence of *On the Equilibrium of Planes*, which Leonardo uses and cites specifically and which he appears to have read in both translations. Also of interest is a long fragment from Book II, Proposition 10, of Moerbeke's translation of *On Floating Bodies* that seems to be in Leonardo's hand (*Codice Atlantico*, 153rb, rc, ve).

38. See Leonardo da Vinci, MS L, 2r.

39. In 1522, at Nürnberg, Johann Werner published a *Commentarius seu Paraphrastica ennaratio in undecim modos conficiendi eius problematis quod cubi duplicatio dicitur* which was stimulated by Valla's translation of the pertinent sections from Eutocius' *Commentary*.

Commentary on the Paper of Marshall Clagett

By Edward Grant, Indiana University

IF WE were asked to select those ancient Greek scientists, mathematicians, and natural philosophers whose subsequent influence on the course and development of science was fundamental and profound, I suspect that our irreducible list would contain the names of Hippocrates, Plato, Aristotle, Euclid, Archimedes, Apollonius, Ptolemy, and Galen. All in this illustrious group, except Galen, have been studied extensively and with considerable care. But when we seek to assess the subsequent influence of these Greek scientific luminaries whose collective works formed the very core of western science as it passed through Byzantine, Arabic, and Latin civilizations, it is astonishing and disappointing how meager is our knowledge.

Few systematic studies have been made on the translations of their works into Arabic and Latin, with the attendant changes in meaning and content that undoubtedly occurred in the process, especially for the large number of works whose path of translation was first from Greek into Arabic and then from Arabic into Latin. For the most part, we also remain ignorant of the manner in which these Greek treatises were later excerpted, sometimes re-titled, and frequently truncated or otherwise mutilated and altered. So profound and complex were the consequences of the process of transmission involving the works of the Greek authors mentioned here, that it is small wonder that few scholars have had the courage, patience, and research capability required before one might dare attempt to unsnarl the tangled mass of linguistic, substantive, and manuscript confusions that accumulated through many centuries.

Only in recent years have scholars sought to determine with exactness the fate of these many Greek works. As an indispensable first move, modern editions must be established of the medieval Latin texts, followed by intensive study of the manner in which they were used and understood. This essential process has thus far begun in earnest only for Aristotle, Plato, and Archimedes. But judging from the prodigious efforts of Professor Clagett, as evidenced by the excellent paper we have just heard, it is a safe conjecture that Archimedes will emerge as the first of the great Greek scientific authors whose subsequent fate and influence will have been thoroughly documented down through the Renaissance.

Professor Clagett has briefly described how the works of Archimedes passed through Byzantine, Islamic, and Latin cultures. We learned that Gerard of Cremona's translation of the *Measurement of the Circle* gave rise to a number of emended versions which testify to a rather persistent Archimedean influence. Coupled with other Archimedean techniques introduced in Gerard of Cremona's translation of the *Discourse of the Sons of Moses* and subsequently supplemented by Johannes de Tinemue's *On Curved Surfaces*, there already existed a considerable body of sophisticated Archimedean geometry prior to the translation of nearly the whole Archimedean corpus from Greek to Latin by William of Moerbeke in 1269. From that year on, much of the best of Greek geometry was potentially available to mathematicians in the Latin West.

Was this precious body of geometric literature of significance in the history of medieval mathematics? Did it encourage and intensify the pursuit of a high level geometry? Was it productive of new theorems and important applications to physical problems?

On the basis of Professor Clagett's account, the overall accomplishment and achievements were rather disappointing. True, a number of medieval mathematicians utilized a few of the Archimedean works with skill and, on occasion, even suggested improvements or alternative methods for justifying or arriving at certain theorems, as did Johannes de Muris in his tract on the *Quadrature of the Circle*. Moreover, Professor Clagett calls de Muris' treatise "the first known commentary on Archimedes' *On Spirals*." And yet throughout the fourteenth and fifteenth centuries no significant medieval Latin commentator or expositor appeared to reveal a general mastery of the works of Archimedes, a fact perhaps partially explicable by a lack of supplementary Greek mathematical works of the kind available to Eutocius in the sixth century A.D.

As for the direct influence of Archimedes, Professor Clagett observes that Nicole Oresme joined "the dynamic definitions of specific gravity" taken from the *Liber de ponderoso et levi*, one of the statical treatises showing the indirect influence of Archimedes, "with the Archimedean considerations of *On Floating Bodies*." With keen and perceptive scholarship he notes further that "it is just such a juxtaposition that appears in Tartaglia's Italian translation and commentary on Book I of *On Floating Bodies*, which may have led Benedetti to his modified form of the Peripatetic law, namely that bodies fall with a speed proportional to the excess in specific weight of the falling body over the medium." This is, of course, the same law of fall which Galileo adopted in *De motu*. Thus Oresme's use of Archimedes heralded a new approach that would be productive of a law of fall in direct conflict with Aristotelian physics. The indirect impact of Archimedes was, however, of far greater moment in medieval physics, for it was instrumental in producing the statical treatises linked with the name of Jordanus of Nemore.

Despite these achievements, the influence of the works of Archimedes in the Middle Ages was relatively modest. Few, if any, new theorems were demonstrated as a result of it. Indeed, it is curious that virtually all of the significant medieval mathematicians such as Leonardo Fibonacci, Jordanus of Nemore, Gerard of Brussels, and perhaps Johannes de Tinemue lived and worked prior to Moerbeke's translations. Moreover, subsequent to those translations, medieval mathematical interests in the fourteenth century shifted markedly to what may be aptly described as the philosophical foundations of mathematics, a subject much more congenial to the philosophically oriented scholastics.

Why did intensive interest in Archimedes have to await the sixteenth and seventeenth centuries when his works captured the attention of men such as Commandino, Guido Ubaldi del Monte, Benedetti, Stevin, Kepler, Galileo, Cavalieri, and many others? In the final analysis, only Professor Clagett will be sufficiently learned in the details of the Archimedean tradition to provide a meaningful answer to this difficult question. But I suspect that one ingredient of that answer will lie in the advent of printing during the 1460's. The role of printing in the history of science has never been carefully assessed, but it could hardly be less than momentous.

Mathematical works as complex as those of Archimedes required accurately drawn and lettered figures, as well as faithful textual reproduction. The errors and omissions which abounded in medieval mathematical manuscripts is amply illustrated in Professor Clagett's first volume on Archimedes. A manuscript in Paris might vary in crucial ways from one in Florence or Oxford. A potentially capable mathematician might have been compelled to invest more intellectual energy in merely comprehending and correcting a corrupt and confusing text than in adding to, or going beyond, it. Even if he were capable of originality, his own works would inevitably suffer some degree of corruption in subsequent copyings.

To generate a continuous, consistent, and cumulative mathematical tradition under such circumstances was virtually impossible. Only printing from movable type could introduce a measure of uniformity of text and figure that would guarantee that readers in Rome, Paris, and Oxford were reacting to identical theorems. With the texts of Archimedes this did not happen until the sixteenth century when Benedetti, Galileo, and a host of others became the beneficiaries of the new technology.

In light of this, the medieval achievement takes on new and deeper dimensions. In the face of insuperable obstacles, they not only translated and preserved the works of Archimedes, but occasionally suggested improvements and applications that marked the beginning of serious Archimedean studies in the Latin West.

It remained for their more fortunate successors, armed with printed editions, to first master and then use Archimedes more effectively and fruitfully and thereby produce a new physics and a higher level of mathematics. But as Professor Clagett has so brilliantly revealed, these printed editions owed much to the medieval tradition. It was, therefore, a fitting tribute to that tradition that the first printed works of Archimedes in 1503 were taken from the Latin translation of William of Moerbeke.

Commentary on the Paper of Marshall Clagett

By John E. Murdoch, Harvard University

It has become something of a habit among twentieth-century medievalists to bespeckle the Middle Ages with suggested Renaissances (and, of course, also to disclaim at least most of them as soon as they have been suggested). It is not my purpose to weigh the merit of this particular preoccupation, but merely to note that in most cases the proposed Renaissance carries the mark of having recaptured a part of at least something from Greek antiquity.

Thus, the twelfth-century Renaissance (easily the most respected among the medieval members of this genre) has as its most important "happening" the translation of an overwhelming number of philosophical and scientific works from the Greek and the Arabic. To know *that* such translating activity took place and *just which* works were turned into Latin is, of course, of utmost significance for our ability to appreciate the magnitude and nature of the "intellectual quickening" constituting the Renaissance. Indeed, together with the rise of the university, this retrieval and absorption of ancient and Islamic material is a foundation that any of us must lay if we are to go on to construct anything within the history of medieval science. It is a foundation, moreover, whose features must be known in considerable detail. We must know, that is, not just which parts of Euclid, Archimedes, Ptolemy, Galen or Aristotle were put into Latin, but exactly what the results of this transfer were.

It is on just this point that Marshall Clagett has, with his first volume of his Archimedes-Latinus, put us immeasurably in his debt. And to judge from the preview which he has given us of the volumes yet to come, clearly we shall soon owe him even more. For the exhaustiveness with which he will cover the medieval Archimedes will, I assure you, long go unchallenged in the scholarship of the translations of the Middle Ages. This, when coupled with the almost excruciating incisiveness with which he has treated the specific details of his task, means that we shall soon have before us not only a complete, but a model, presentation and analysis of medieval mathematical source material.

Yet having said this, let me shift my stance—for at least a moment —and play the role of *advocatus diaboli*. Surely, we all readily admit, the painstaking edition of medieval translations is required if we are to have *their* Archimedes, *their* Euclid, or *their* Aristotle, and not merely the ones classical philologists have given us. Still, admitting this, when it comes to reading through—even cursorily—all these translations, not a few of us might feel inclined (albeit unconsciously) to whisper: Is this really necessary? Must I really wade through all of this properly to appreciate medieval science and mathematics?

If, in reply, I put aside the fact that it would do no harm, the most telling point is that without directly addressing oneself to such material, a considerable portion of the essence of medieval science would be missed. For—and this is the basic point of my whole comment—the *complete* presentation of the medieval versions of the works of Greek mathematics tells us not only what the nature of these versions were, but a good deal about the nature of the scholastic mathematical enterprise as well. I shall, at least for purposes of argument, even go so far as to say we can learn as much, if not more, about scholastic mathematics from these translations as from original mathematical works written in the Latin Middle Ages.

Let me support this (seemingly outrageous) claim by beginning with the medieval Euclid and working my way back toward Clagett's Archimedes. First, a few facts by way of orientation. (1) The Euclid of which I shall be speaking, is the Arabic-Latin *Elements* (the Greek-Latin versions were either fragmentary or ignored in the Middle Ages). (2) Within this Arabic-Latin tradition there are over twenty different versions of the *Elements*, most of them related to a core of versions established by Adelard of Bath and Campanus of Novara. I shall draw basically on this core. (3) Most of these Latin versions of the *Elements* are just that, versions, and not, strictly speaking, translations. This is due basically to the fact that the proofs given of the propositions often diverge (frequently radically) from what one has in the Greek text.

Given all this, what do we find in this particular medieval Euclid, and what can it tell us of the general tenor of scholastic mathematics? We have, to begin with, what is basically a "teacher's" Euclid. Didacticism is everywhere present. We are constantly reminded of the nature of a proof and of the appropriate axioms and previous propositions being utilized within it. Postulates and axioms are even added to fill every possible gap. Illustrative material is introduced to clarify the complex, and *exempla secundum numeros* are employed to facilitate the comprehension of geometrical assertions.

Directly connected to, indeed part of, this pedogogical tone is an ever-present emphasis upon the logic of the *Elements* and upon the role of foundational notions within it and within mathematics in general. Thus, great care and attention is paid to definitions, postulates, and axioms and to the most basic propositions, which means not only the propositions that are appealed to most frequently within the course of the *Elements*, but, more significantly, those with a heavier bearing upon philosophical issues within mathematics and natural philosophy. And this focusing on logic and the fundamental becomes even more apparent when one considers not just the comments of Adelard or Campanus (and their followers) in the text of the medieval version at hand, but the often extensive marginalia— marginalia that tell us of the use to which the scholastic put his Euclid and of what of value he saw in him. At times the concentration in both the text and marginalia upon foundational notions is so great as to suggest that there was more concern with such notions than with the mathematics itself. This would, admittedly, be fittingly scholastic.

If we return now to Archimedes, it is apparent that at least a part of his medieval phase bears precisely those characteristics we have noted of the Adelard-Campanus Euclid. That part is the *De mensura circuli* in all of its numerous Arabic-Latin versions. It, too, exhibits the didactic, troubles to spell out just which more elementary mathematics is being appealed to in its proofs, emphasizes logical structure, and reveals a concern for the foundational in its specification of added axioms (mostly continuity postulates) that are tacitly relied upon.

What is more, the subject of which this Archimedean opusculum treats, the quadrature of the circle, was itself, more than any other segment of this Hellenistic mathematician, of considerable philosophical interest to the scholastic. It was Aristotle, far more than Archimedes, who made this so. His citation of the problem of quadrature in his logical works and in his *Physics* made the entrance of something more of the mathematics of the problem standard scholastic procedure. Medieval expositors find the problem most relevant even when Aristotle had not mentioned it, but had cited only such related issues as the comparison of rectilinear and curvilinear motion.

Moreover, the Aristotelian context underscored the already-present emphasis on axioms or fundamental notions, since there the interest was frequently in the logical relation of quadrature (so those of Antiphon, Bryson and Hippocrates of Chios) to geometrical principles. Going even a step further, the scholastic felt that Aristotle's manner of viewing the possibility of circle quadrature (*si est scibile, scientia quidem eius nondum est*) made it pertinent fare for the logical treatises called *Obligationes* (in which one was concerned with the precise role of assumptions or concessions granted within an argument or debate).

It seems, however, that the treatise on quadrature by the fourteenth-century Albert of Saxony is more characteristic of the medieval expression of the philosophical bearing of the problem. It, together with the elaborated medieval versions of the *De mensura circuli* to which it refers, would appear far more in tune with the late scholastic mathematical appetite than would a straightforward (William of Moerbeke) rendering of Archimedes' Greek without frills. At least this much is said when, in 1390, a German Franciscan saw fit to copy Albert's work *plus* another variant version of the *De mensura circuli* into a geometrical compilation which he expressly says is designed *ad introductionem iuvenum in geometriam*.

The quadrature of the circle, then, as a fundamental mathematical problem with appreciable philosophical relevance, furnished grounds for the most frequent citation of Archimedes and, as a corollary, for the most widespread significance he held for late medieval science and natural philosophy. This, and not the more sophisticated weighing of his work by one like Johannes de Muris, is, I would like to suggest, more characteristic of what was going on in fourteenth-century mathematics.

To draw an historian's moral, then, we should realize that the kind of toiling Marshall Clagett has done, and is doing, in setting forth the full spectrum of variant translations of a figure like Archimedes is far more than the mere assembling of "dry bones." Such work reveals not just the exact form Archimedes or Euclid had for their medieval consumers, but also what their works meant to them, what was thought of them, and why they were considered relevant and useful. The various elements that, as we have seen, drew special attention in the medieval versions of Archimedes and Euclid furnish a basis, I feel, for answering such questions. At the same time they tell us of the nature of late scholastic mathematical thought. They tell us that it was highly philosophical.

Viewed from another perspective, this is to say that a substantial sector of mathematical endeavor in the fourteenth century did not have to do with the extension or development of the central concerns of Greek mathematics. There was, for example, precious little of the reworking and broadening of Greek techniques of quadrature and cubature that one finds in Islamic mathematics. On the contrary, the scholastic mathematics of which I feel we have evidence in the medieval versions of Euclid and (at least the most popular work) of Archimedes is one which arose, not from an attempt to extend this inherited mathematics, but rather from a constant effort to relate it to logical and philosophical issues, to pick out and to expand anything in this mathematics of the least significance to such issues. This is what I mean by calling it philosophical.

But this was not the only strain of philosophical mathematics in the fourteenth century. There was another that arose, as it were, in an inverse direction. It did not render an existing mathematics philosophical; rather it grew out of philosophy. It proceeded, to be specific, from a kind of fourteenth-century frenzy to measure as many of the variables within philosophy as possible (and within theology as well).

Penes quid attenditur was the war-cry for the attack. The strategy was to begin by establishing some (presumably inviolate) basic rule for the measure of that being measured: speed measured by the fastest moving point, uniformly difform qualities measured by their mean degrees; the perfection of a species measured by its distance from *summum esse* (or, if you like, backwards from *non esse*), or (to use our terms) an arithmetic variation in speeds measured by a geometric variation in the ratio of their determinant causal forces and resistances. Shades, of course, of Bradwardine, Heytesbury, Swineshead, Oresme and all the others. The point is, however, that although a modicum of mathematics was employed in the establishment of these rules (and sometimes a bit more to offer proofs of them), the most substantial mathematical development sets in when the scholastic begins to test the mettle of these rules through the imagination of all conceivable complications and "fringe" cases.

Consider, for instance, bodies undergoing rarefaction and condensation or even partial corruption (rotating wheels of ice in an oven), limit the rarefaction to every 2^{nth} proportional part and have each such part rarefy twice as slowly as the preceding one. Drop a rod through a hole through the earth to determine if it will reach the center of the universe. From such unusual cases applied to the basic rules of measure, new mathematical techniques or abilities arise. We are witness, for example, to what *in our terms* amounts to the establishment of convergence and divergence for infinite series, and to the conception and manipulation of irrational exponents.

But all of this is also not an extension or development of Greek mathematics proper. The analogue of the scholastic natural philosopher in the fourteenth century plying his *scientia mathematica de motu* is not, before him, Euclid or Archimedes, or, after him, Galileo or Descartes. He is, rather, medieval through and through. He is directly akin to the medieval philosopher in his often indefatigable "testing" of a basic rule of measure; our late scholastic was but acting in standard fashion. For he needed but to look about him to see the same procedure being applied everywhere within philosophy. All definitions, axioms or theories were treated to the same kind of ordeal by special or unusual test cases. His stretching, shrinking mobiles might be replaced by infinite golden mountains and goat-stags, and instead of mathematics there may have been something more like metaphysics, but the genre of analysis was all but identical.

Bradwardine, Swineshead, and Oresme were, then, basically doing philosophy, but doing it mathematically. The kind of mathematics that resulted from their activity, together with the kind of philosophical mathematics one can glean from the medieval reaction to Euclid and Archimedes, yield what is most characteristic, I should like to suggest, of late scholastic mathematics as a whole. All this needs, of course, a great deal more study, and the work Marshall Clagett is doing will here prove of infinite help.

Stephen Moulton Babcock—Benevolent Skeptic

By Aaron J. Ihde, University of Wisconsin

BOOMING LAUGHTER frequently traveled down the stair well of South Hall, startling Dean William Henry in his ground floor office and causing him to wonder what Babcock was up to now. The disturbance was a repetitious one, but the dean had finally given up warning his new chemist to consider the dignity of his professorship in the agricultural college. More commonly now, he was tempted to climb the four flights of stairs to learn what could be so humorous about a scientific experiment.

Stephen Moulton Babcock was a lighthearted man. In an obituary statement, University of Wisconsin President Glenn Frank characterized him as the "Laughing saint of science." In further remarks Frank said, "This merry man . . . was made of the stuff that gives mankind its saints and martyrs. But he was a saint without seriousness, and he could have gone to martyrdom, without a murmur of self-pity, as part of the day's work. . . . He pursued the most painstaking research as if he were playing a game. . . . He did not think it impertinent to doubt the authorities . . . he was an adventurer into the unknown to whom research was an intellectual ritual."[1]

A humble man, Babcock never took himself seriously despite the honors heaped upon him by a grateful dairy world. Nor did he take science seriously. He would sooner munch peanuts at a baseball game than listen attentively to a profound research paper at a scientific meeting. He published little. He felt relieved when others took his most profound ideas and examined them experimentally.

Babcock was born on a farm near Bridgewater, New York, on October 22, 1843. Following a typical upbringing on a sheep farm he became a student at Tufts College and earned the A.B. degree in

Presented as the Rosetta Briegel Barton Lecture at the University of Oklahoma on April 10, 1969. Acknowledgment is due the National Science Foundation for support of this investigation.

271

1866. At that point his interests turned to engineering and he matriculated at Rensselaer Polytechnic Institute. His studies, however, were cut short when his father's death necessitated a return to Bridgewater to undertake operation of the farm.[2]

The family farm provided an unstimulating life for this young man with a curious mind. Farm life posed innumerable questions, but his searching mind was out of step with the drudgery of farm operations, and he soon associated himself with Professor G. C. Caldwell at Cornell University, where he served as a chemistry assistant. In 1875 he was made an instructor in the subject, but two years later he left the post in order to undertake graduate studies in chemistry at the University of Göttingen in Germany. This chemistry department will be remembered as the one headed for four decades by the illustrious Friedrich Wöhler. Wöhler was now nearing the end of his life and no longer taught graduate students. Babcock received his Ph.D. in 1879 under Hans Hübner.

Upon returning to the United States, Babcock resumed his instructorship at Cornell, but left it in 1882 to become a chemist at the Geneva branch of the New York Agricultural Experiment Station. It was here that he worked on a gravimetric method for analysis of milk for fat, studied the size and number of fat globules in milk, and devised a viscosimeter for determination of adulteration of fats and oils. His work at the station was competent, but not particularly distinguished. It was typical of the kind of work that was being done at the time in the new agricultural experiment stations throughout the United States.

In 1888, at age 45, Babcock left his position at Geneva to become Professor of Agricultural Chemistry and Chief Chemist at the Agricultural Experiment Station at the University of Wisconsin. This position had been held during the previous five years by Dr. Henry Prentiss Armsby, an authority on animal feeding. Armsby left Wisconsin in order to become director of the agricultural experiment station in Pennsylvania, where he carried on distinguished work in energy balance in farm animals.

At this particular time, dairying was undergoing a major transition. In the past, farm families frequently owned a few cows which were milked to supply the household needs. In the season of high production, a surplus might be churned into butter or converted into cheese for local sale. However, there had just been developing a trend toward factory production of butter and cheese which was stimulating local farmers to increase their dairy herds and sell milk as a cash crop.

Since milk was sold by the hundredweight, a serious problem was developing. The honest farmer was in competition with farmers who unscrupulously added water to the milk in order to increase their income. Still others skimmed cream from the top of the milk before delivery to the cheese factory and turned the cream into butter for home use or as a source of secondary income.

Such adulteration of milk was difficult to detect with certainty. The lactometer, a specially designed hydrometer with a scale calibrated to the specific gravity of milk, could be used to detect watering because the specific gravity of milk is approximately 1.032, compared to 1.000 for water. The lactometer was also suitable for detection of skimming since milk fat, having a density of about 0.9 is the lightest portion of milk. When milk has been skimmed its density becomes greater than 1.032. The lactometer could, however, provide no certain proof against adulteration when the farmer both skimmed and watered the same batch of milk. Thus, he was able to gain in both directions while leaving the factory operator with no sure evidence of unethical manipulation.

The fat content of milk is a good indication of quality, but all of the fat tests in use in 1888 were tedious and generally required the services of a trained chemist. Scientists in various experiment stations in the United States and Europe had been seeking simplified methods for determination of the fat content of milk, but none of these modified tests was sufficiently rapid and reliable for commercial use.

When Babcock joined the Station at Wisconsin, Dean Henry soon assigned him the task of designing a suitable method for determination of fat in milk and cream. Despite his previous experience with methods for milk analysis, and despite his knowledge that others had been unsuccessful in devising a satisfactory test, Babcock set to work seeking to modify the well-known Soxhlet method for extraction of fat from foods by means of ether, with subsequent evaporation of the ether and weighing of the fat left as a residue. As a consequence of diligent developmental work, Babcock soon devised a test which proved successful for the determination of butterfat in composite samples of dairy herd milk. Being skeptical of easy success, he then sought to verify the reliability of the test by checking it on milk samples from each of the cows in the University's herd. Results compared favorably with those obtained by the slow but accurate official method for fat determination until Babcock encountered the milk of a grade Jersey named Sylvia. Sylvia's milk gave results at variance with those obtained by the official method.

Dean Henry urged that Babcock's method be published despite failure to test accurately the milk of one of the cows, arguing that cheese makers and creamery operators would be using the test on composite samples of herd milk delivered to the factory, but would seldom be testing the milk of single cows. Babcock refused, abandoned the partially successful test, and struck off in new directions.

He was aware that various investigators had utilized strong acids or alkalies to release milk fat from its normal suspension. By trying various ratios of concentrated sulfuric acid and milk he soon settled on a volume of acid which, by charring the milk sugar and protein, released the fat without damage. The fat could then be concentrated at the top of the mixture by centrifuging and brought up into the calibrated neck of a specially designed test bottle by dilution with hot water. Once in the neck the quantity of fat could be measured directly in percentage by means of forceps. The test was so simple that it could be completed in a few minutes by scientifically unsophisticated persons. The method was published in June, 1890.[3] Babcock insisted that he have no remuneration for the test, but that it be released for the free use by the dairy world.[4]

The test proved a boon to the dairy industry. Its simplicity made it possible for factory operators to run fat tests periodically on composite samples of milk delivered by their farm patrons. Payment might then be made on the basis of butterfat delivered rather than on the basis of pounds of (sometimes adulterated) milk delivered. The farmer received payment for a quality product and the cheese maker was now in a position to anticipate a standard yield from normal milk which filled his vats. Watering and skimming were abandoned since such practices were no longer profitable. The editor of *Hoard's Dairyman* stated that Babcock's test was a greater influence than the Bible in making farmers honest.

The Babcock test also played a role in improvement of dairy cattle. Testing associations in dairy states had as their purpose the evaluation of butterfat production by individual cattle in tested herds. Thus, it was possible for farmers to eliminate unprofitable animals (star boarders) and to breed for quality milk production.

Despite the many honors which were heaped upon him, Babcock tended to depreciate his role as the savior of dairying. Upon occasion he pointed out that he merely developed ideas which had been tried previously, but never carried out to the point where a quick practical test resulted. Once the test was developed he took comparatively little interest in extending its application. Such work was done by lesser associates in the experiment station, such as Fritz Woll, E. H. Farrington and J. L. Sammis. Through minor modifications of the test bottle and the methodology it was possible to apply the test to cream, skimmed milk, whey, and other dairy products.[5]

Babcock's next investigations were carried out in collaboration with Dr. Harry Russell, the station bacteriologist. The pasteurization process was beginning to come into widespread commercial use in order to enhance the keeping quality of commercial fluid milk and make it free of disease producing organisms. When the pasteurization process was extended to other dairy products such as cream, consumer complaints became extensive. Pasteurized cream failed to whip satisfactorily. Babcock and Russell investigated the whipping quality of pasteurized cream and showed that pasteurization caused a breakdown of fat globule clusters, thus impairing whippability. They found that whipping quality might be restored by the addition of calcium sucrate, a product obtained by treating cane sugar with quicklime.

Skeptical about the current pasteurization practices which exposed cream to very high temperatures for long periods of time, they showed that the chemical additive was quite unnecessary. Adequate destruction of disease-producing bacteria occurred when cream was pasteurized at moderate temperatures for short time periods, while whipping quality was not destroyed.

The unique collaboration between chemist and bacteriologist continued as the pair turned their attention to the curing of cheese. At the turn of the century, when factory-made cheese had become an important product of the dairy industry, the cheese was frequently of such poor quality that it was rushed to market as rapidly as possible. The consumer, therefore, was purchasing cheese with underdeveloped flavor resulting from its youngness, or cheese with harsh and unpleasant flavors developed during the fast curing process. Babcock and Russell showed that normal curing of cheddar cheese was brought about by an enzyme normally present in milk as well as enzymes added with the rennet used for coagulation of the milk. If cheese were held at a temperature below 60°F., they showed that there was a marked improvement of flavor accompanied by reduced loss of moisture as compared to the customary method of curing at environmental temperatures. The cold-curing of cheese quickly became standard in the dairy industry. By questioning customary practices in the industry Babcock and Russell brought about a superior product which benefited the industry significantly.[6]

Although Babcock was closely identified with the milk industry his background, interests, and associations caused him to be aware of the broad variety of agricultural problems which might in some way be related to science.

There was much interest at this time in the scientific value of foods and feeds. If agricultural experiment stations were to prove their value they must clearly reveal to farmers how they might obtain high production with a minimum of feed input. If one looks at agricultural experiment station reports during the last decades of the nineteenth century and on into the twentieth, one finds principal attention given to such matters as the kind of feeds which will produce most rapid growth, greatest production of wool, milk, or eggs, greatest production of power in a horse, and matters of this sort.

Virtually all agricultural investigators who were interested in feed values were followers of the beliefs which had been laid down by the German chemist Justus von Liebig at the middle of the nineteenth century. Liebig was of the opinion that two ingredients were essential in foods; energy producing materials and a source of nitrogenous food (protein) for replacement of tissues destroyed during metabolism. Liebig was also of the opinion that proteins were composed of a standard radical carrying characteristic amounts of sulphur and phosphorus.

The measurement of energy values of foods was easily undertaken. In this period when thermodynamics was undergoing rapid development by physicists, nutrition scientists eagerly jumped on the energy bandwagon and measured the energy values of foods. Respiration calorimeters were developed from the middle of the nineteenth century and respiration calorimetry was the most popular activity in nutrition studies toward the end of the century. Stanley Benedict at Connecticut Wesleyan and Henry Armsby at Penn State College were building respiration calorimeters large enough for human beings and for farm animals. Benedict even built one large enough to study the metabolism of an elephant. It was quickly recognized that carbohydrates and fats are solely sources of energy.

Proteins also serve as an energy source, but are also important for rebuilding of tissues. Since proteinaceous foods tend to be more expensive than those noted solely for their energy value, there was a marked tendency among animal feeding experts and experts in human nutrition to design diets in which emphasis was placed upon cheaper food materials such as cereal grains. Feeds high in proteins were only included to the extent necessary to provide the nitrogen necessary for replacement of that which was used in production of work, wool, eggs and milk.

Shortly before the end of the century, C. W. Langworthy of the U.S. Department of Agriculture published analytical tables of human foods not only bringing out the chemical analysis of such foods, but emphasizing the importance of spending one's food dollar for cereal grains which are high in energy while holding meats, milk, fruit, and vegetables at a minimum level. It was pointed out in the Langworthy tables that in consuming foods which are high in water, as is the case with milk, fruits and vegetables, the consumer is really wasting his food dollar because he gets much less caloric value than he does in cereal grains.

Although he was educated in Germany where food analysis was developed extensively at an early date, Babcock always remained skeptical of the importance of chemical analysis of foods. He was particularly skeptical about equivalencies of different foods and frequently raised embarrassing questions about the equivalency of proteins from different sources. He was aware that Armsby, his predecessor at Wisconsin, had been the author of a book on animal feeding where these matters were treated in a highly scientific manner. He was aware of the large amount of analytical work which was going on in experiment stations all over the world, including the one at Wisconsin where Henry was sponsoring extensive analyses of various feed materials. William Henry himself would ultimately become the author of a widely read work on feeds and feeding. Despite his friendship for Henry, and his proximity to feed analyses, Babcock remained unimpressed.

While he was still at the Geneva Station he had carried out extensive analyses of dairy cattle feeds. One day he placed before his chief, Dr. Sturtevant, two analytical reports, querying Sturtevant regarding which would be the better ration for a dairy cow. Sturtevant ran his eyes over the analytical figures carefully before venturing a conclusion that there seemed to be very little difference between them. With a twinkle in his eye, Babcock ventured the opinion that this seemed very strange since one analysis represented the feed which went into the cow, the other was the excrement which came out.

At meetings of the Association of Official Agricultural Chemists Babcock frequently plagued his colleagues with questions about their faith in chemical analyses. Once he suggested that it would be possible to design a ration from purely mineral sources which would contain all necessary ingredients for a satisfactory animal ration. When challenged, Babcock suggested they look at the composition of bituminous coal. The nitrogen multiplied by 6.25 might be reported as protein. Material extractable with ether might be reported as fat. In addition, coal contained materials reportable as crude fiber and nitrogen-free extract. Presumably, coal might therefore take its place in rations recommended as animal feeds.

On another occasion Babcock inquired why leather might not be ground up and used as cattle feed, since it was rich in nitrogenous material. Despite such tendencies toward *reductio ad absurdum*, Babcock's contemporaries continued to believe that feeds might be shuffled around in animal rations as long as proper attention was given to total protein and energy.

At around the turn of the century Babcock had an opportunity to test his ideas regarding the importance of biological testing of feeds. He secured the loan of two cows, one of which was placed upon a balanced ration coming entirely from the oat plant and the other on a balanced ration coming entirely from the corn plant. Both animals soon showed obvious deterioration and when the oat-fed animal died, Professor Carlyle of the Animal Husbandry Department withdrew the other animal from further experimentation. Animal husbandry experts in those days were more concerned about the sleek appearance of their animals than about letting them be used for experiments which could potentially be hazardous to their well-being, even though such experiments might result in improved knowledge of feeding.

Growth and changes in the agricultural college led to a well-designed repetition of the single grain feeding experiment starting in 1907. E. B. Hart, a young chemist from the New York station, had now joined the agricultural chemistry department, and George Humphry had become director of the animal husbandry department when Professor Carlyle took a position elsewhere. Harry Steenbock, a hard-working farm boy who was still an undergraduate, became a member of the group. In order to have the services of a talented analyst, E. V. McCollum, a Yale Ph.D., was hired by Hart.

Babcock had no part in running the experiment, even though the basic idea of a long-term feeding experiment on a restricted but presumably chemically-adequate diet was his. He had a deep interest in the results, but he was now busy seeking the answer to puzzling questions regarding metabolic water.

In setting up the single-grain experiment, Professor Hart, a superb experimenter, utilized sixteen cows over a four-year period. Starting out as heifer calves, these animals grew to maturity and went through two gestations. Four of the calves were placed upon a "scientifically" balanced ration made up entirely of parts of the wheat plant, four others received a similar ration derived from the oat plant, another four were fed on the corn plant, and the last four were fed a ration made up of equal parts of the three single-grain diets. This latter group was intended to serve as a scientific control. All of the diets contained equivalent amounts of protein, energy sources, and minerals.

After the experiment had been carried on for six months the wheat-fed animals showed marked deterioration, and such deterioration soon became evident in the oat-fed animals and those fed on the mixed ration. Only the corn-fed animals were vigorous and retained a good appearance. It was obvious at this point that chemically equivalent rations derived from different plant species were not biologically equivalent.

Upon reaching maturity the animals were bred and went through a first gestation. The wheat-fed heifers were unsuccessful in giving birth to healthy calves. One was born dead, two died during the first day, one lived twelve days. The oat-fed animals gave birth to only two calves that lived, and even these were not healthy. The animals on the mixed diet gave similar results. Only the corn-fed animals appeared healthy and gave birth to healthy calves. The animals were bred a second time and the results were similar, except that by now two of the wheat fed cows were dead and neither of the two remaining gave birth to a calf capable of survival. During the last year of the experiment the surviving cows were shifted to another ration (wheat to corn, corn to wheat, wheat to oats, etc.). Cows shifted to the corn ration quickly showed improvement; corn-fed cows shifted to wheat or oats quickly deteriorated.

The report published in 1911 was looked upon by the experiment station with great pride, since it revealed clearly the lack of biological equivalency in feeds even though chemical equivalency was maintained.[7]

In the meantime, McCollum despaired of this kind of experimentation. He quickly became convinced that chemical analysis was incapable of revealing the causes of biological deficiencies in foods. As an ardent student of the literature, he soon unearthed a dozen cases where health had been unsatisfactory when humans or animals were fed diets from a highly restricted source or materials in a high state of purity. McCollum soon concluded that the cow experiments represented a poor way to explore the problem. He said as much to Professor Hart, who chided him for so quickly losing interest in an experiment which presumably would have the greatest possible significance in the field of dairy husbandry.

One Sunday morning late in 1907, McCollum was at work when Babcock strolled into the laboratory and sat down to visit. The two men had become fast friends, the elderly Babcock finding a great stimulus in the young man from Yale who was well read in the chemical literature and seemed to be bubbling with new ideas.

McCollum revealed his discouragement with the cattle experiment, but indicated that he must keep on because of the pride which the station was taking in the studies. Babcock asked him how he would prefer to experiment. McCollum answered, "I think we should use rats instead of cows. Rats don't eat much. They have a short life span. Presumably through biological testing we might find the answers to some of these questions which chemical analyses are incapable of answering." Babcock was quickly persuaded of the sagacity of the young chemist. He told him, "I happened to see Dean Russell coming up the street as I entered the building: Let's go down and talk to him."

Babcock and McCollum turned up in the Dean's office a few minutes later. Babcock reported that McCollum had an idea worth hearing. McCollum proceeded to tell about his discouragement with the cattle experiments and his conclusion that feeding experiments should be carried out on rats. Russell rose from his chair, pounded the desk and roared, "The rat is a barnyard pest and should be exterminated. The legislature will never stand for feeding such animals. We must proceed with cattle."[8]

A few days later Babcock appeared at McCollum's workbench and, with a twinkle in his eye, suggested that McCollum proceed with his rat feeding ideas. Babcock said, "I had a talk with Professor Hart and he has no objections and the Dean needn't know about this. There is plenty of scrap feed around the experiment station so such experiments shouldn't be costly."

With this encouragement McCollum proceeded to trap wild rats in the university stables and began his rat feeding experiments. These wild rats, however, proved to be highly intractable and utterly unsuitable for animal experimentation. Shortly thereafter, McCollum purchased 12 albino rats from a pet dealer in Chicago, paying $2 for the animals. At the time of his death he had not been reimbursed for these rats, which were the beginning of the Wisconsin colony.

The rest of this history is quite well known. McCollum soon found that rats grew unsatisfactorily upon diets of highly purified proteins, carbohydrates, fats, and minerals, but that the addition to the ration of small amounts of milk resulted in healthy growth. The same observation was made almost concurrently by T. B. Osborne and L. B. Mendel at the Connecticut Agricultural Experiment Station. Mc-

Collum went on to demonstrate the presence of accessory food factors (Vitamins A and B) in butterfat and milk solids.

Babcock's name did not appear on the publication which resulted from the single-grain experiments, nor did it ever appear on any of McCollum's papers. Yet there is no question that it was Babcock's skepticism coupled with an agricultural experiment station environment which opened the door to the newer knowledge of nutrition.

Babcock retired from his position as professor of Agricultural Chemistry upon reaching seventy years of age in 1913. He had just published the results of his lengthy studies on metabolic water. These, typically, arose out of his whimsical questioning of nature. He was aware that all living creatures contain an abundance of water. Why then, can certain ones, such as clothes moths, beeswax moths, pea weevils, and confused flour beetles, survive on dry foods with no external source of water? Babcock found the answer in their use of metabolic water produced by the oxidation of dry food materials, and associated with biological mechanisms for the conservation of such moisture.[9]

Babcock's retirement years were spent in Madison, where he and his wife lived in their modest home a few blocks from the university. His passion for hollyhocks occupied much time during the growing seasons. He learned to drive an automobile during his eighth decade of life and enjoyed touring rural Wisconsin with Mrs. Babcock. But he was regularly seen during this period making the trek to the Agricultural Chemistry Building where he enjoyed the company of younger colleagues while pursuing studies on the nature of matter. He had developed skepticism regarding traditional views on gravity. This work was never published. After his death on July 1, 1931, his notebooks were examined by several physicists who concluded that there was nothing of significance therein.

The career of Stephen Moulton Babcock presents many enigmas. His scientific career developed slowly. At the time he came to Madison at age forty-five, he had done scientific work of only a pedestrian character. The test which brought him worldwide fame was merely patient development of approaches used unsuccessfully by others. He himself depreciated the test and looked upon the studies of metabolic water as his most significant work. His searching skepticism about food equivalencies led others to uncover the role of trace components in nutrition.

Babcock can hardly be looked upon as a scientist of the first rank. Yet, through his perseverance, his willingness to team up with members of other disciplines, and his good-natured resistance toward current paradigms, he set into motion investigations which were to have profound implications in agriculture and medicine.

1. Glenn Frank in Harry L. Russell, *et al.*, "Stephen Moulton Babcock . . . A Memorial to him in Observance of the Centenary of His Birth" (Wis. Alumni Research Foundation, Madison, 1943), pp. 25–27.

2. On Babcock's life see A. J. Ihde in Eduard Farber, ed., *Famous Chemists* (Interscience, New York, 1961), pp. 808–13 and the bibliography on pp. 828–29.

3. S. M. Babcock, "A New Method for the Estimation of Fat in Milk, Especially Adapted to Creameries and Cheese Factories," *Bull. Univ. Wis. Expt. Sta.*, no. 24 (1890).

4. The test worked satisfactorily on milk of individual cows, including Sylvia's, as well as on herd samples. It was never ascertained why Sylvia's milk gave erroneous results when using the earlier test.

Sylvia came to an ignominious end a few years later when the tuberculin test was run on all cows in the university herd. The herd was found to be badly infested with the disease, almost all animals reacting positively. It was decided to slaughter all of the cows. Autopsies confirmed the test results.

5. J. L. Sammis, "The Story of the Babcock Test," *Circular* 172, *Extension Services, College of Agr., Univ. of Wis.* (Madison, 1924).

6. S. M. Babcock and H. L. Russell, "The Cold Curing of Cheese," *Wis. Agr. Expt. Sta. Bull.* 94 (1902); Edward H. Beardsley, *Harry L. Russell and Agricultural Science in Wisconsin* (Univ. of Wisconsin Press, Madison, 1969), pp. 49–63.

7. E. B. Hart, E. V. McCollum, H. Steenbock and G. C. Humphrey, "Physiological Effect on Growth and Reproduction of Rations Balanced from Restricted Sources," *Wis. Agr. Expt. Sta., Research Bull.* 17 (1911).

8. As told to the author by E. V. McCollum on April 9, 1964. Also see E. V. McCollum, *From Kansas Farm Boy to Scientist* (Univ. of Kansas Press, Lawrence, 1964), pp. 118–19.

9. S. M. Babcock, "Metabolic Water: Its Production and Role in Vital Phenomena," *Wis. Agr. Expt. Sta., Research Bull.* 22 (1912).

The Science of History and the History of Science

By Joseph T. Clark, S.J., Canisius College

INSTANT communication which is truly transparent is no more readily to be expected, even between academic strangers who would become friends, than a cup of instant coffee anywhere which is truly flavorful and indescribably delicious. Some brewing and percolating time is required beforehand in order to establish a common idiom of expression and a mutual understanding of key concepts. I therefore respectfully request your indulgence for a brief spell, while we reconnoitre together some landmarks of importance in the intellectual territory which we all hope ultimately to share in joint possession.

I should like first of all to make certain that we all comprehend exactly what is meant by a "protocol statement" such as philosophers understand to be a standard item of procedure in natural science. A typical protocol statement is a formal report which contains, for the most part, the following kinds of data: space coordinates, time coordinates, specifying circumstantial details, and then a bald description of the phenomenon then and there observed. Most specimens also close with the signature or initials of the observer, as an affidavit for its authenticity. A simple but highly instructive example is to be found in a sample temperature report which a ward nurse enters into hospital records: Bed No. 12 (space coordinates), 8 April 1969, 7:43 P.M. (time coordinates), Mr. John Johnson (subject observed), oral application (specifying detail), temperature reading: 98.2° (the observed phenomenon).

I should like also to make certain that we all comprehend exactly what is intended by the contemporary science of semiotic and, in particular, the roles played therein by the correlated concepts of "functor" and "argument" in the logical, as distinct from the merely grammatical, syntax of language. Let us focus briefly for illustrative

Presented at a meeting of Phi Alpha Theta at the University of Oklahoma, on April 8, 1969.

purposes on two notorious instances of syntactical nonsense uncovered by semiotic analysis in otherwise highly respected sources in the history of philosophy. The first example pronounces pontifically in a highly sonorous context that *das Sein ist identisch*. But semiotic observes that *identisch* is a two-place functor and makes syntactical sense if and only if it occurs in association with a pair of arguments, such as in the authentic case: "Cicero is identical with Tully." The second specimen asserts with metaphysical profundity that *das Nichts nichtet*. But semiotic remarks that "nothing" is not a genuine argument in the syntactical category of names for which "nichtet" could serve as a one-place functor, but rather a mischievously misleading syncopation of ordinary language to express the negation of a universally quantified expression. But negation is syntactically a functor and never an argument. Hence the second sentence, constructed of functors only, fails like the first to achieve the minimal semiotic status of good syntactical sense.

I should next like briefly to state that in contemporary idiom "an axiom system" does not mean a system of true theorems deduced from a prior set of absolute, necessary, and infallible axioms, such as our grandfathers used to think was the case with the *Elements* of Euclid. Rather, an axiom system is currently understood to be a nonempty set of sentences S, such that S is exhaustively partitioned into two and only two subsets K and K', and such that every member sentence of S is either a member of K or a member of K', but not of both, and such that K' contains all but only those sentences of S which are derivable by appropriate applications of conventional rules of inference from some or all of the member sentences in K. To endorse this axiom system S is neither to affirm the independent truth status of the member sentences of K nor to assert the detached truth status of the theorem sentences of K', but only to affirm the conditional sentence: "If K, then K'."

I should further like to distinguish effectively between "explicit definition" and "implicit definition." In an explicit definition a new term is defined by means of terms priorly present in the systematic vocabulary. But it is impossible to define all terms explicitly. Hence the meaning of some terms can only be recognized by observing the contexts in which they are employed. Euclid, for instance, with highly questionable success attempted to define "straight line" explicitly as "a line which lies evenly with the points on itself," rather than implicitly as "a line which satisfies the requirements stipulated in the accompanying postulates." And, in general, the same observations pertain to the relatively primitive terms of any axiom system in any field whatsoever.

I should like also to distinguish effectively between "knowledge by acquaintance" with a given factual situation, and "knowledge by understanding" of the same empirical complex. This distinction is basic to all talk about "explanation," historical or otherwise. Let us assume that we both *know* by empirical encounter that something S is in a condition P. To *understand* how it happens that S is P, we seek what we call an *explanation* of the phenomenon. If one attempts to solve our problem by showing (1) that all S is M and (2) that all M is P, it is generally agreed that a satisfactory explanation has been offered for the circumstance that S is P. In short, explanation of a problematic sentence S consists fundamentally in engrafting S via logical connections onto a priorly established and accepted axiom system A.

I should also like, finally, to establish effectively a crucial distinction between a "law of logic" and a "rule of inference." This distinction is crucial despite the fact that each involves the form of a conditional compound sentence: an antecedent sentence usually introduced by the particle "if," and a consequent sentence conventionally introduced by the paired particle "then." A simple illustration of such a conditional compound sentence is the following: "If it is raining, then the streets are wet," wherein the antecedent sentence states a sufficient condition for wet streets, namely, rain; and the consequent sentence announces a necessary condition for the occurrence of rain, namely, wet streets. Such a conditional compound sentence is certainly true when it is the case both that it is raining and that the streets are wet. And such a conditional compound sentence is certainly false when it is the case that it is raining, indeed, but it is *not* the case that the streets are wet. But such a conditional is also to be construed as true when it is *not* the case that it is raining but it is, nevertheless, the case in point of fact that the streets are wet. For the conditional sentence in question does not claim that rain is *both* a sufficient *and* necessary condition for wet streets; it does not state that the streets are wet if and only if it is raining. Such a conditional, finally, is also to be reckoned as true when it is both *not* the case that it is raining and *not* the case that the streets are wet. For neither circumstance was therein stated categorically to be a fact, but only the second, *if* the first. In brief: a conditional compound sentence is truth-functionally false if and only if its antecedent sentence is true and its consequent sentence is conjointly false.

But there are some such conditional compound sentences which are always true. These sentences are indefeasibly valid, and constitute laws of logic. Let the literal "p" stand as dummy and surrogate for any declarative sentence that you wish. And let the literal "q" stand as companion dummy surrogate for any other declarative sentence that you wish. If so, then the compound conditional sentence which reads:

$$p \cdot p \supset q \cdot \supset q \qquad (1)$$

can never ever be false but is always indefeasibly true or valid. For it turns out that whenever the consequent sentence is false, so too in each and every case is the antecedent sentence.

But it is as important as it is infrequent to observe that such laws of logic are *not* rules of inference. A law of logic is a compound conditional sentence which merely states via dummy surrogates a brute fact of inert validity, whereas a rule of inference authorizes and warrants a thinker to do something about it, namely, to come into rightful possession of a new and true sentence q. The law of logic, for example, previously instanced:

$$p \cdot p \supset q \cdot \supset q \qquad (1)$$

is the ground for the infallible rule of inference known in the jargon of the ancient craft of logic as *modus ponendo ponens*, or the independent affirmation of the antecedent sentence p. So that if there is warrant for the assertion that wet streets truly accompany rain, and one is further in a position to certify that it is in fact now raining, his stock of known true sentences may now legitimately be increased by detaching the consequent sentence and knowing for sure, without empirical inspection at first hand, that the streets are infallibly wet.

This distinction between a law of logic and a rule of inference is not only of great theoretical interest to the career logician. Else I should never have mentioned it in this context. This distinction is also the source of that single invaluable insight which alone makes decisively important methodological researches into the rational structure of the historical enterprise for the first time a feasible undertaking of serious scholarship. And nothing, I take it, could be more pertinent to our present discourse. For this distinction makes mandatory the bifurcation of all types of human reasoning into two and only two broad classes: not deduction and *in*duction, since then history would be neither and thus remain pathetically nondescript, but rather deduction and *re*duction.

This analytical result means, in effect, that each and every instance of explanatory proof, wherever it occurs, in science or in history, is reducible to a pair of paradigm premises, so arranged that the first is a conditional sentence: "If p, then q," and the second is either p or q. The two cases may be schematized respectively as follows:

(1)	(2)
If p, then q.	If p, then q.
But p.	But q.
Therefore q.	Therefore p.

An inference which reproduces the structure of schema (1) is deduction. An inference which reproduces the structure of schema (2) is reduction.

The rule of inference which guides the thought process of schema (1) is the *modus ponendo ponens*, or the independent affirmation of the antecedent sentence. This rule is solidly based upon a valid law of logic and is therefore indubitably infallible in each instance of its appropriate application.

The rule, however, which guides the thought process of schema (2) may at first blush appear highly suspect. It is well known in logical quarters that independent affirmation of the consequent sentence is *not* a source of validity. But—as Galileo was perhaps the first to observe—to say that some given form of compound conditional sentence is not true in *all* cases is not equivalent to a declaration that it is false in each and every case. For what is not true in all cases may still be true in surprisingly many cases. And it is high time that more career logicians took into professional account the unquestionable and challenging fact that the frequency index of the incidence of such true cases is astronomically high—or else there would be in fact no historical development of science, nor any science at all—because such reductive reasoning is the very life of every science.

Reduction may further conveniently be subdivided in two significant and vitally correlated ways: (1) predominantly *re*gressive reduction, and (2) predominantly *pro*gressive reduction.

In predominantly regressive reduction one takes as preferable point of departure the experientially known and established truth in contemporary context of the puzzling or problematic consequent sentence, and then moves creatively in thought toward the sophisticated construction of an as yet tentative and provisional antecedent sentence with which the former can be connected as a logically necessary consequence. Here, precisely, is the logical locus of the celebrated raw data or impartially observed matters of fact with which every science inaugurates each new phase in its historical and helical sequence of suitable solutions to authenticated problems. Such predominantly regressive reduction is also the rational anatomy of the procedures of thought heretofore known to classical philosophers of science as theory construction, the logic of discovery, or explanation of a given set of phenomena.

In predominantly progressive reduction one takes as preferable point of departure the as yet tentative and provisionally constructed antecedent sentence, and then moves in thought toward a decisive confrontation in experience with the contextual content of a predictable consequent sentence, connected with the former as a further logically necessary consequence. Here, precisely, is the logical locus for consciously contrived experimentation in the natural sciences. And such progressive reduction is also exactly the rational anatomy of the thought processes heretofore known to classical philosophers of science as verification or confirmation, and their opposites: falsification or disconfirmation.

A further, and as the sequel will undertake to disclose, highly important distinction in reduction arises out of the quantificational index attached to the antecedent sentence. If the antecedent sentence is in effect a generalization of the contents of the consequent sentence, then the reduction reduces for all practical purposes to induction. If such, however, is not the case, then it is important to specify the reductive process as *non*-inductive.

What, then, *is* the science of history?

It is conventional in even otherwise informed circles to suppose that the crucial differences between the enterprise of the natural sciences and the science of history are chiefly three.

The first difference has long since been immortalized by the nineteenth-century German historians in finely chiseled phrases to the effect that whereas the natural sciences concern themselves with *die Welt als Natur*, the science of history explores *die Welt als Geist*.

The second difference is said to lie in the fact that whereas natural science establishes laws which transcend temporal specifications, such as the achronic equation of Einstein: $E = mc^2$, the science of history explores the past, precisely as past.

The third alleged difference is that whereas the physical sciences venture to *explain* to a literate public just how phenomenal events predictably occur in nature, history assiduously and ascetically eschews explanation of any kind at all to anyone at all and conscientiously restricts its entire enterprise to reporting, reliably indeed, but merely describing with disciplined phenomenological reserve *wie es eigentlich gewesen ist*.

But it seems to me that the first alleged bifurcation between *Natur* and *Geist* cannot survive serious scrutiny. For the *Geist* of man is not disembodied and transempirical and angelic, but incarnated in throbbing three-dimensional flesh and inextricably entwined and intermeshed with the material matrix of *Natur*.

The second purported difference fares no better under critical analysis. For each and every phenomenon which experimental science explores is technically indeed a *past* event, if often of the more recent past, whereas the science of history chooses by professional preference to explore events of a more remote past. But this distinction, even if relevant at all, would be at most a difference of degree, not in kind, and therefore both trivial and negligible.

The third conventional difference pivots upon the historian's alleged abdication of any or all explanatory endeavors and is so enormous a prevarication as to be in violent contradiction with the overt behavior of the historical fraternity, at least since the time of Thucydides. For history does indeed venture to explain its phenomena. No historian ever was or ever will be able to resist indefinitely the itch to explain the events of which he has by his researches acquired expert knowledge. All historians undertake to explain: the good ones very well, and the bad ones very badly. But the pretended professional moratorium on explanation itself is, I think, readily explicable. Torturously confronted in the nineteenth century by an intolerable, but false, dichotomy between deduction and induction—neither of which historians could with a clean conscience accept—the severely descriptivist myth was cozily invented and professionally propagandized as a desperation maneuver to ensure in serious crisis at least the mere survival of a venerable discipline in the republic of letters.

If these three are thus fraudulent, there are nevertheless some genuine differences between the natural sciences and the science of history, and it will profit our purpose to detect and to define them. I detect at least two, the second, furthermore, genetically connected with the first. The first methodological difference is that whereas history, like science, undertakes to explain events, it does so without the spectacularly enriching resource of inductive procedures. The second difference is also methodological in character and filially related to the first in so far as the absence of inductive processes forecloses for history the possibility of uttering significant and reliable generalizations.

Despite these genuine differences, it is my confident view that the science of history is an empirical science, sister, sibling, kith and kin, or, at the very least, kissing-cousin to the notoriously and flamboyantly empirical sciences of nature. For the raw materials of history are reportorial sentences about phenomena, here too precisely construed in the authentic scientific sense of observable events. The fact that these phenomena appropriate to history are not of the immediate or more recent past is irrelevant and immaterial in the context. For not only is a similar situation imaginable in the natural sciences, it is an on-going reality in paleontology, geology, and astro-physics.

Nevertheless, it is undeniable that this professional predilection for events of the remoter past dooms the historian to contend constantly with an immeasurably greater degree of complexity in the execution of his reductive craft. And it is not with the intent of green-eyed envy that I remind experienced historians how the practicing scientist has at his ready disposal in the heretofore unread, perhaps, archives of his discipline a prodigiously enormous reserve supply of protocol statements, each expertly couched in a technically precise idiom by trained researcher colleagues who share with their fellows a cultural milieu so uniform and homogeneous that there is in principle no obstacle to instant and authentic and complete understanding. It is with the explicit purpose of evoking empathy, if not sympathy, that I also remind experts in the craft how the historian is compelled by the professional demands of his trade to begin his labors with what are termed euphemistically, and euphorically, "documents." For it is notorious how profoundly such documents differ from the protocol statements of science. These raw materials of history, even when not also written in a language which is not the mother tongue of the researcher, are for the most part haphazard residues of a cultural complex weirdly and eerily unfamiliar and strange to the researcher himself. Behind the syntactical façade of the words in such documents there lurk most often the ghosts and shades and hobgoblins of a completely alien conceptual scheme characteristic of the cultural epoch selected for historical investigation.

It is clear, therefore, that if history possesses its protocol statements as science does, these do not lie ready to hand for the historian, but must be uncovered by a protracted and tedious and extremely difficult process of reconstruction, decipherment, exegesis, and interpretation. Then and only then can one make reliable statements about the mere facts in the case. The sciences of nature and the science of history thus both exhibit with equal clarity the twin methodological phases of (a) statements concerning individual phenomena, and (b) statements of an explanatory nature. But in history alone we find a distinctly different third phase of operation, prior to the point in natural science where protocol statements come into play. This is the phase, preliminary to (a), where interpretation labors strenuously to construe authentically the sense of the documents at its disposal. The logical phase sequence for history is therefore threefold: (1) documentary processing, (2) the protocol-like statements of ascertained fact, and then (3) statements of frankly explanatory import.

In this context it is also sometimes said that a third and authentic difference between the science of history and the natural sciences is to the effect that whereas the historian is always faced with an initial problem of selection and choice, the scientist is not. I do not think that this observation accords with the facts. It is indeed the case that the sheer mass and welter of documentary materials is so vast that one of the first responsibilities of the historian is the awesome task of determining a wise choice and selection among them. But is it not obvious on the face of it that the scientist constantly confronts an even richer set of protocol statements and has before him an incredibly greater total of observable phenomena open to investigation? Far more things happen in the universe than happen to become documented for a future historian's use. The difference does not lie in the fact that the conscience of the historian must confront the grim necessity of an initial option, while the scientist does not. For in point of fact the pinching shoe is on the other foot.

But to all outside appearances the scientist who wears the shoe hurts not, whereas the historian—sometimes dramatically and in the public gaze to boot—agonizes over his predicament. Why is this paradoxically so? The difference lies, I suggest, in the fact that whereas the scientist possesses an impersonal criterion of selection, the historian does not. Confronting a far greater superabundance of materials, the scientist chooses more simply, more easily, almost automatically and by instinct than does the historian, because the reductive procedures of natural science are *also* inductive. What interests the scientist exclusively is that and only that portion of the circumambient data that is subject to the generalizing procedures of inductive methodology.

If not by an impersonal criterion, such as induction in the case of the scientist, then by what standard, if any, does the career historian professionally determine his selection? To this genuinely disturbing question I am presently convinced that no decisive answer can ever be forthcoming—and on principle. For each attempted reply inevitably invites a recurrence of the same as yet unanswered question. If one contends, for example, that the initial selection of a suitable research problem automatically predetermines all subsequent selections of relevant materials, the respondent must then expect to be asked by what criterion the initial selection of that research problem was itself determined. And so *ad infinitum*.

I candidly confess that I see no other solution to this problem of selection in history than to admit publicly that when all else is said and done by way of defense or pretense, a subjectively warranted value-judgment is decisive in each case. That, at any rate, is how I became an historian of science, and not of politics or economics or warfare. I find in this admission no cause for shame or embarrassment or apology. Personal choice on the basis of one's own scale of values for one's own reasons and from one's own motives enters only at the point where the phenomena to be investigated are definitively selected. Once this decision has been effectively made, admittedly on subjective grounds, the professional research enterprises of the historical investigator are no less objective and impartial and impersonal than those undertaken through the compulsive pressures of inductive possibilities in the area of the natural sciences.

It is, I think, no digression at this point to distinguish very clearly between the literary problem of history and the pseudo-problem of literary history. For whereas there is no such animal in my zoo as literary history, there is indeed a very genuine literary problem of history. And that problem here merits a brief reexamination and, if possible, a more accurate redefinition.

Even a cursory review of the standard journal reviews will establish that an extremely high value is currently placed by the profession upon the degree of excellence in literary style exhibited by a historian's published works. This perverse and persistent prejudice puzzles me profoundly. For when one analytically penetrates beneath the polished surface veneer of stylistic graces, and scrutinizes with a sober eye the actual residue of intellectual content therein expressed, it always seems to come as a surprise and sometimes as a shock to the authors to learn that, like Molière's M. Jourdain, they have subconsciously been striving to speak in the idiom of the leaner, more sinewy, and more muscular prose of the exact sciences all the time.

If this diagnosis is correct—and the evidence for it will soon be forthcoming—then the real literary problem of history is not by what various new and other stylistic devices and stratagems more effectively to compete with professedly fictional "historical" creations for public attention and support, but rather how to become, consciously and deliberately, less "literary" and more "logical" in presentation.

I have previously listed the logical phase sequence for history in these three stages: (1) documentary processing, (2) protocol-like statements of ascertained fact, and then (3) statements of frankly explanatory import. Let us agree to employ this triple schema for marshalling in economic fashion the evidence promised to support the foregoing diagnosis of the literary problem of history.

First of all, it is quite remarkable to me to note how the success of phase one, documentary processing and research, crucially depends upon the semiotic and axiomatic techniques expertly developed and employed in the exact sciences of logic and mathematics, but to date exploited with far less rigor and exactness by historians. Here lies in relative neglect a powerful tool for the improvement of our trade.

In phase two, where protocol statements are being formed and fashioned, success is a function of the adroit employment of the rules for implicit definitions within the structure of axiomatic systems, long since soundly established and rigorously utilized by practitioners of the exact sciences, but to date exploited only informally and somewhat casually by the craftsmen of historical research. Here lies in relative neglect a second powerful tool for the advancement of our trade.

Again in phase three the procedures to be followed are in no serious way different from the familiar procedures of scientific explanation, expertly executed and routinely so by scientific personnel trained to the task, but heretofore used by historians only incidentally and without conscious attention. In each case one explains a problematic sentence S by enmeshing it via logical connections into the structure of an axiomatic system A. But it stands to reason that axiom systems which are consciously constructed are superior to those which are uncritically inherited from one's predecessors in the trade, or passively absorbed by a sort of unconscious osmosis from the prevailing *Zeitgeist*. Here, too, lies in relative neglect a powerful tool for the improvement of our profession.

At the completion of phase two the historian is for the first time sufficiently equipped to venture upon the specific phase three task of historical explanation. But from this moment on, no notable difference in the rational structure of the thought processes required of both the historian and the scientist can be discerned, except of course the degree of currently superior quality of performance and execution of trained personnel in the respective sciences. Exactly as the research scientist is long accustomed to proceed, so, too, must the historian consciously seek by reductive methods to explain his newly ascertained protocol statements of fact by engrafting them onto a logical framework of other accepted statements. Like the scientist, he must consciously further exploit the many other available regressive reduction and contrapuntal verification procedures. Here lies in relative neglect a fourth powerful tool for the advancement of our professional fraternity.

There are thus, as I see it, no critically major differences in the rational structure of the thought processes productive of explanations in the natural sciences and in the science of history respectively. But there are some minor differences of detail, some worthy of a brief mention.

The first is to the effect, simple to state but enormous in its consequences, that historical science is devoid of all opportunities for experimental research, precisely because of its concern for single past events, forever gone with the wind and unreproducible in identical or even comparable contexts. Thus the science of history must exclude from the inventory of its reductive resources even the primitive variation methods of Mill, for example, as well as other established and more ingenious techniques of experimentation, easily available to the natural sciences, even at the relatively low level of clinical medicine. Here, I fear, if anywhere, is to be found the root source of the relative —but also irremediable—incompleteness of the historical enterprise.

The second difference is to the effect that historical explanation, at best, is almost always genetic in character. A problematic sentence A is explained by means of a sentence B, the significant terms of which refer to events prior to the time dimensions ingredient in sentence A. Then the same historian proceeds to explain the reconstrued problematic sentence B by means of a sentence C, the significant terms of which refer to events prior to the time coordinates of the sentence B. And so on and on throughout the continuity of the historical process. Such genetic explanations are not peculiar to history, indeed, but they play therein a more dominant role than anywhere else.

But such minor differences do not obscure for me the central identification of the operative thought processes in both the natural sciences and the science of history. Emphatically, *both are empirical sciences*. Neither is deductive in apparatus. Each is reductive, but whereas science is also inductive, history is not.

But if such be the intellectual structure of the scientific enterprise in itself, and if such be the rational anatomy of the science of history in general, what, then, are we to say in particular about the specialized discipline of the history of *science*?

I think it a fair reply to answer that the history of science is plainly and simply history. Not military history, of course, nor political history, either, nor diplomatic history, nor economic history, nor social history, nor religious history—not even just cultural or intellectual history *überhaupt*. Rather, within the conglomerate context of all the foregoing sister disciplines, it is the history precisely of the mental itinerary of those specific ideas which are known—by present 20–20 hindsight—to have been at work in shaping contemporary man's evolutionary conceptual scheme of his own world and his systematically developmental understanding of its presumed structure and its environmental processes.

Where, for example, the general historian with a military bent chooses to reconstruct for himself and then to explain to us the rise of Roman civilization and its subsequent predatory expansion throughout the period of the three Punic wars, I prefer to pay attention to the roughly contemporary scientific work of Archimedes, Hipparchus, Euclid, Eratosthenes, and Apollonius of Perga.

And where, for example, the general historian of religious trend undertakes to reconstruct for himself and then to explain to us the origins of the Protestant Reformation and its subsequent vicious vicissitudes through to the end of the Thirty Years War, I prefer to pay attention to the roughly contemporary scientific work of Tycho Brahe, Simon Stevin, Gilbert, Galileo, Kepler, Mersenne, Descartes, Fermat, Snell, Harvey, and Torricelli.

And where, for example, the general historian with political preferences volunteers to reconstruct for himself and then to explain to us the emergence of the Commonwealth in England and the ensuing violent vicissitudes of Britain through to the time of William of Orange, I prefer to pay attention to the roughly contemporary scientific work of Christiaan Huygens, Isaac Newton, Robert Hooke, and Robert Boyle.

And where, for example, the general historian with diplomatic interests opts to reconstruct for himself and then to explain to us the genesis of the American Revolution and its oppressive engagements thereafter through to the establishment of the United States Constitution, I prefer to pay attention to the roughly contemporary scientific work of Lagrange, Laplace, Black, Fourier, Watt, Franklin, Cavendish, Coulomb, Volta, and Lavoisier.

And where, for example, the general historian with economic interests elects to reconstruct for himself and then to explain to us the emergence and the rapacious expansion of the industrial revolution in Europe through the first quarter of the nineteenth century, I prefer to pay attention to the roughly contemporary scientific work of Carnot, Oersted, Ampère, Ohm, Herschel, Young, Fresnel, Davy, Dalton, and the immortal Faraday.

And where, for example, the general historian with social anxieties determines to reconstruct for himself and then to explain to us the eruption of socialistic movements and their ensuing fratricidal embroilments from the time, say, of the Communist Manifesto by Marx and Engels to the outbreak of the Crimean War, I prefer to pay attention to the roughly contemporary scientific work of Mayer, Joule, Kelvin, Clausius, Galois, Cayley, Wöhler, Lyell, and Schwann.

And why?

Because I find, after fifty as I did before fifteen, that *all* such scientific ideas—whether successful or not, whether fertile or not—rest gentle on my mind, and that their intriguing history reveals to me— so much more authentically, I believe, than the records of wars or charters or treaties or gross national products or class conflicts or even creeds—what I take to be the true statural dimensions of the observable phenomenon of man.

It could, therefore, just possibly be the case, as the late George Sarton was wont to suggest, that the history of *science* is, in fact and in deed, the *new* humanism for our contemporary, irreversibly technological, and at the present moment miserably beleaguered culture.

INDEX

Absolute: change of motion, 201; force, 182, 200; motion, 183, 188, 199–200, 203–205; rotation, 187, 201; space, 183, 187, 201, 204–205; system of measurement, 75; time, 187
Acceleration: 183, 186, 200–201, 207
Accessory food factors: 281
Acid: 43, 63–64, 142–43
Actualism: 212, 221–22, 224–25, 230, 234, 236
Adelard of Bath: 265–66
Aesthetics: 129–76
Affinity: 47–48, 59, 63
Agassiz, Louis: 4, 11, 20, 225
Agricola, Georg: 149
Agricultural experiment stations: 272, 276–82
Agrimensores: 245
Air: 44–45, 47–48, 56
Aitchison, Leslie: 159
Albert of Saxony: 244–45, 257, 267; *Questio de quadratura circuli*, 244–45
Alberti, Leon Battista: 250, 258; *De ludi matematici*, 258
Al-Bīrūnī: 243
Alchemy: 61, 147, 149
Alembert, Jean Lerond d': 200
Alexander, Gustav: 163
Alexander, Shirley: 160
Alexis [pseudonym]: 162
Alhazen: 242; *Optics*, 246
Alps: 233, 236
Al-Tūsī, Nasīr al-Din: 241
American Association of State Highway Officials: 121
American Society of Civil Engineers: 119
Amman, Jost: 149, 163–64; *Book of Trades*, 149
Andes: 219
Anthemius of Tralles: 240
Anti-mechanistic views: 46, 72, 74, 78–79
Antiperistasis: 178
Anti-reductionism: 59–60
Antiphon: 267
Antonius de Albertis: 253

Aqua: fortis, 41, 43; *martis*, 143; *regia*, 41, 43
Apollonius of Perga: 260, 295
Arabs: 239, 241, 260–61
Arbor Dianae: 145
Archimedes: 239–69; *Cattle Problem*, 240, 248; *De mensura circula*, 245, 266; *Lemmata*, 242, 254; *Liber assumptorum*, 242; *On Conoids and Spheroids*, 249–50, 256; *On Data*, 242, 254; *On Floating Bodies*, 240–41, 248, 250–53, 256–57, 259, 261–62; *On Parallel Lines*, 242, 254; *On Spiral Lines*, 242, 251; *On Spirals*, 241, 248–50, 256, 261; *On the Equilibrium of Planes*, 240–42, 250, 254, 256, 259; *On the Division of the Circle into Seven Equal Parts*, 242; *On the Measurement of the Circle*, 240–41, 243, 245, 247, 249–52, 256, 261; *On the Method of Mechanical Theorems*, 240–41, 248; *On the Properties of the Right Triangle*, 242–43; *On the Quadrature of the Parabola*, 241–42; *On the Sphere and the Cylinder*, 240–41, 247, 249–51, 253, 255–56; *On Touching Circles*, 242; *On Triangles*, 242; *On Water Clocks*, 242; *Opera*, 252; principle of, 250; *Sandreckoner*, 248, 251; *Stomachion*, 240–41, 248
Archytas: 246
Aristotle: 4, 8, 10, 32, 51, 61, 65, 260–61, 264–67; *Physics*, 266
Armor: 142, 153
Armsby, Henry P.: 272, 276–77
Arrhenius, Svante August: 72–73
Art: 129–76; affected by science, 151–53; and communication, 172–73; and physical forces, 156–57; and symbolism, 153–55, 170; and technical innovations, 168; close relations with technology of materials, 167; defined, 132; interaction with technology and science, 129–76; relationship between materials and techniques, 132–33; technical side of, 155; used to extend human experience, 152; Western trend away from representational, 156

The paper on which this book was printed bears the watermark of the University of Oklahoma Press and has an effective life of at least three hundred years.

UNIVERSITY OF OKLAHOMA PRESS

NORMAN